LINEAR ELASTIC ANALYSIS

LINEAR ELASTIC ANALYSIS

DAVID G. ELMS

Senior Lecturer in Civil Engineering
University of Canterbury, Christchurch, New Zealand

B. T. BATSFORD LTD *London*

First published 1970

0 7134 0513 9

Made and printed in Great Britain by
William Clowes and Sons Ltd, London and Beccles
for the publishers B. T. Batsford Ltd
4 Fitzhardinge Street, London W1

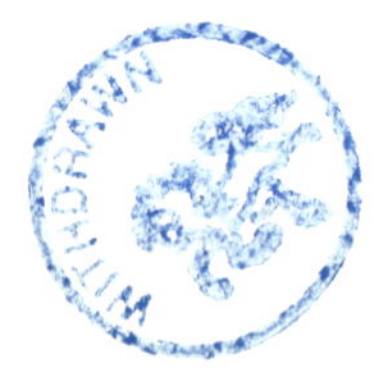

Contents

Preface

Elastic structural analysis is a logical subject. Only a relatively few standard formulae need be used and remembered, for the main requirements of the subject are clear thinking and a thorough understanding of the principles involved. This book therefore has two main aims; it tries to convey:

1. An understanding of basic principles rather than specific techniques, and
2. A 'feel' for the behaviour of structures.

Specific analysis techniques are, of course, dealt with in some detail, but they should be as much regarded as illustrations of particular aspects of the general force and displacement approaches to elastic structural analysis as thought of as techniques to be mastered in their own right; for a knowledge only of current techniques will not give an engineer the versatility he must have in order to be able to adapt himself to new methods as they arise. In this book I have therefore tried to give an integrated and cohesive treatment of the basic principles of structural analysis of linearly elastic systems.

All ideas have been illustrated by applying them to frame structures. I have not dealt with more specialised structures such as arches or suspension bridges, and neither have I considered the more complex types of solution that would be involved in the analysis of plates or thin shells, for instance, or in the use of finite element techniques. Matrix notation is used throughout the book wherever I have thought it helpful, but it has never been used as an end in itself except, perhaps, in the final chapter.

As a prerequisite for the book the student should know about

1. The principles of statics and the analysis of statically determinate structures.
2. Linear elasticity and the principles of superposition.
3. Bending moment and shear force diagrams for beams.
4. Methods for finding the deflection of beams.

All these are revised briefly in the first chapter. An elementary knowledge of matrix algebra is also necessary.

Primarily in order to avoid confusion due to the change from the Imperial to the S.I. system of units, all numerical examples and problems have been written without specific dimensions. Any consistent set of units can therefore be assumed to apply.

I wish to express my sincere thanks to Professor Robert Park, whose suggestion initiated the project and whose comments and encouragement helped to carry it through; to Professor David Billington who was the source of many of the ideas; to Professor H. J. Hopkins and my other colleagues at Canterbury, whose continued interest was a spur; to the many students on whom various ways of teaching the subject, some fruitful and some later discarded, were tried; and to Dorothy Ball for her patience in typing the manuscript.

1

Revision: Equilibrium, Kinematics and Elasticity

1.1 Forces and Moments

1.1.1 Newtonian Equilibrium

The use of equilibrium equations in terms of forces and moments is fundamental to all structural analysis. For any free body which is not accelerating, the sum of the components of all forces in any direction is zero, and the total moment about any axis is zero. For a general three-dimensional body these statements lead to six independent equilibrium equations: three force equations and three moment equations. In vector notation these may be expressed as

$$\left.\begin{aligned} \sum \mathbf{F} &= \mathbf{0} \\ \sum \mathbf{M} + \sum \mathbf{F} \times \mathbf{r} &= \mathbf{0} \end{aligned}\right\} \tag{1.1}$$

where **F** and **M** are any force and moment on the body, **r** is a position vector from some origin and the sums include all forces or all moments. These equations are, of course, only particular cases of the more general equations of dynamic equilibrium.

The concept of a *free body* is fundamental in setting up equilibrium equations: it is a body imagined to be totally disconnected from any of its surrounding supports or structure and acted on by forces or moments. It may be rigid, it may be elastic, it may consist of several parts interconnected by hinges, springs or bolts; but as a whole it must be imagined to be freed from its surrounding environment and supports, whose presence is acknowledged only by the forces and moments they apply to the structure.

Any structure can be divided into an infinite number of free bodies, the choice depending on the decision of the analyst. For the highway bridge of fig. 1.1, three possible free bodies are shown in (a), (b) and (c). (a) is the complete bridge with its supports replaced by reactive forces, while (b) and (c) are sections of the bridge with forces and bending moments applied to the cut ends to represent the effects of the parts of the structure which have been removed.

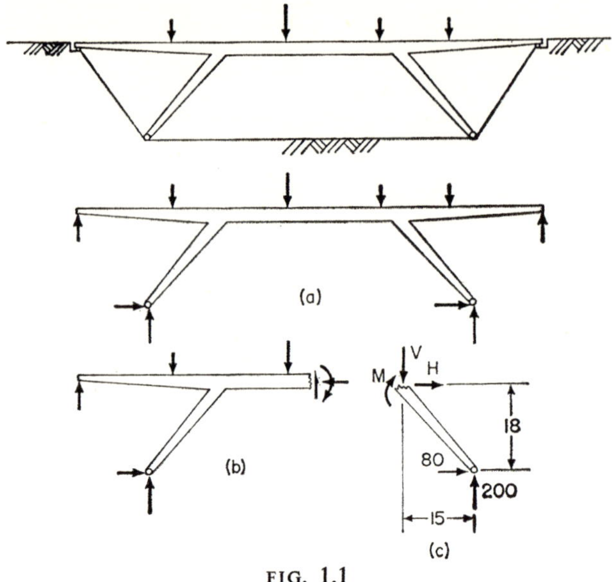

FIG. 1.1

Whichever free body has been chosen, equilibrium equations may be written for it. Fig. 1.1(c) is a two-dimensional example for which three independent equilibrium equations can be written. They are:

Summation of horizontal forces: $80 + H = 0$
Summation of vertical forces: $200 - V = 0$
Summation of moments about the top: $M - 18 \times 80 - 15 \times 200 = 0$

The importance of the concept of the free body will become apparent later, for wherever more complicated statical relationships are encountered in the book, they are resolved by the choice and the use of simple free bodies with their accompanying equilibrium relationships.

1.1.2 Equilibrium by Virtual Work

Equilibrium equations for a structure can alternatively be set up using the principle of virtual work, also known as the principle of virtual displacements for this type of application. The principle states that

> If a rigid body or system of bodies acted on by a system of external forces† is given a small (virtual) set of displacements, then if the forces are in

† The word 'forces' here is used in a generalised way to include moments as well as linear forces. Similarly, the word 'displacements' is used to denote both linear and rotational displacements.

equilibrium the total work done by the forces moving through the virtual displacements is zero.

This could be written

$$\sum \mathbf{F}.\delta\mathbf{\Delta} + \sum \mathbf{M}.\delta\boldsymbol{\theta} = \mathbf{0} \tag{1.2}$$

where $\delta\mathbf{\Delta}$ and $\delta\boldsymbol{\theta}$ represent the small variations in linear and rotational displacements applied to the system. The principle of virtual displacements can also be extended to apply to elastic bodies, but this development will be left to Chapter 7.

It is often much easier to form equilibrium equations using virtual work than to set them up by a straight application of the principles of statics. This is firstly because a virtual work equation does not include forces which do not move and hence do no work, so that support reactions will not generally be included in the equilibrium equation; and secondly because the use of virtual displacements essentially transforms an equation of statics into an equation of kinematics (geometry), for it often happens that the kinematics of a structure is easier to formulate than its statics.† It should be emphasised, however, that both virtual work and a direct application of the principles of statics will ultimately produce exactly the same equilibrium equations, the only difference occurring not in the equations themselves but in their derivation. The following examples illustrate the use of virtual work: the results can of course be checked by statics.

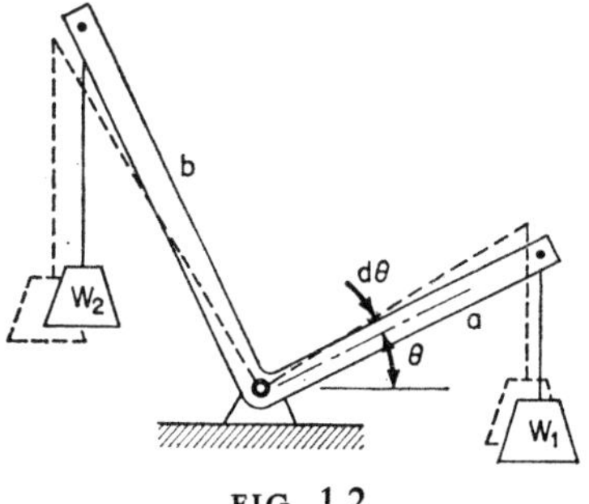

FIG. 1.2

To find the angle θ for which the mechanism of fig. 1.2 is in equilibrium, neglecting the weight of the arms, a small rotation $\delta\theta$ from the equilibrium position is first applied. For equilibrium, the work done is zero, so that

$$W_1(a\,\delta\theta)\cos\theta - W_2(b\,\delta\theta)\sin\theta = 0$$
$$\therefore\; W_1 a\cos\theta = W_2 b\sin\theta$$

and hence

$$\theta = \tan^{-1}\left(\frac{W_1 a}{W_2 b}\right)$$

† It is worth while noting that virtual work can conversely be used to transform a kinematic problem into a problem of statics.

Next, it is required to find the force in bar AB of the truss shown in fig. 1.3(a). The principle of virtual displacements as we have formulated it applies to external forces on a system, and we are now required to find an internal force. To overcome this difficulty and extend the application of the principle to cover internal forces, the bar AB is cut and equal and opposite 'external' forces F are applied to the cut ends. This strategy obviously leaves the overall distribution of forces in the truss unchanged. A small displacement $\delta\Delta$ is then applied to each end of the cut bar, as shown in fig. 1.3(b). It is a simple matter of geometry to show that point C will then move down by an amount $4\delta\Delta/\sqrt{3}$. Hence the work done by the applied forces is

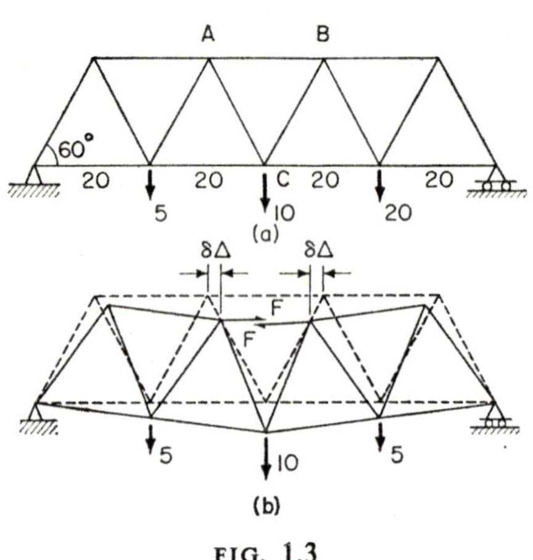

FIG. 1.3

$$\left(\frac{5}{2} + 10 + \frac{20}{2}\right)\frac{4\delta\Delta}{\sqrt{3}}$$

The work done by the force in AB is $2F\,\delta\Delta$, so that for equilibrium the virtual work equation becomes

$$2F\,\delta\Delta + \left(\frac{5}{2} + 10 + \frac{20}{2}\right)\frac{4\delta\Delta}{\sqrt{3}} = 0$$

$$\therefore\ F = -\frac{45}{\sqrt{3}} = -26{\cdot}0$$

Note that the magnitude of the support reactions did not have to be calculated.

The bending moment at A in the beam AB shown in fig. 1.4 may also be found using virtual work. Again the problem arises that we want to find an internal force, whereas the statement of the principle of virtual displacements given in equation (1.2) applies only to external forces, so once more the strategy is used of making M_A an external force. Fig. 1.4(b) shows this done,

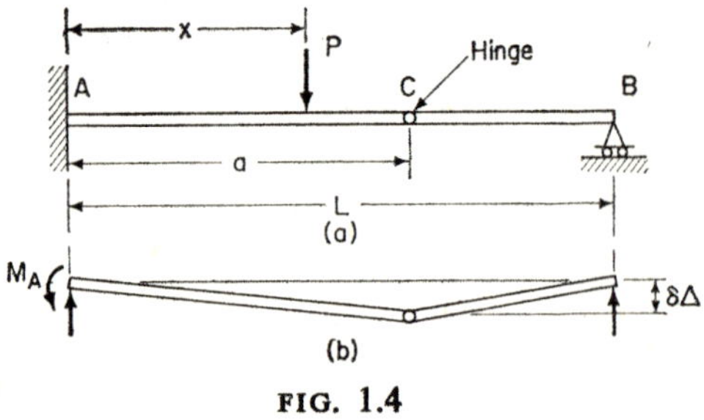

FIG. 1.4

and it also shows the displaced shape of the beam. In fact, taking AB as a free body and remembering that the virtual displacement we apply is purely imaginery and does not occur in the actual structure so that our choice of displacement is not restricted to a permissible displacement of the actual structure, we could apply any displacement system whatsoever. However, the vertical reactions at A and B are not required, so a displacement system is chosen for which A and B do not move vertically and hence do not occur in the equilibrium equation.

If a displacement $\delta\Delta$ is given to the hinge, the work done by the moment at A is $-M_A(\delta\Delta/a)$. Note the sign, which is important. The virtual work equations then become

$$\text{(a)}\quad 0 \leqslant x \leqslant a \qquad -\frac{M_A\,\delta\Delta}{a} + \frac{Px}{a}\,\delta\Delta = 0$$

$$\therefore\; M_A = Px$$

$$\text{(b)}\quad a \leqslant x \leqslant L \qquad -\frac{M_A\,\delta\Delta}{a} + \frac{P(L-x)}{(L-a)}\,\delta\Delta = 0$$

$$\therefore\; M_A = \frac{Pa(L-x)}{(L-a)}$$

1.1.3 Shear Force and Bending Moment Diagrams

If a beam subjected to various loads and reactions is cut at some point, then to maintain equilibrium of the beam a transverse force and a moment must be applied across the cut. These are called the shear force and the bending moment in the beam at that point. A sign convention for shear force and bending moment is seldom used in this book; but where it does appear it is that the shear forces acting on a small element of the beam are positive if they tend to turn it clockwise, and the bending moment at a point is positive if the beam sags at that point.

Shear force and bending moment diagrams are graphs of the shear force and bending moment plotted along a beam. A bending moment diagram is always plotted on the tension side of the member—a convention which is very helpful in general two-dimensional structures, for if a member is not horizontal, 'sagging' has no meaning. The shear force V at a point is related to the bending moment by the equation

$$V = \frac{\mathrm{d}M}{\mathrm{d}x} \tag{1.3}$$

that is, the shear force is the slope of the bending moment diagram. Similarly, the distributed load w on the beam is related to the shear force and bending moment by the equations

$$w = -\frac{dV}{dx} = -\frac{d^2M}{dx^2} \tag{1.4}$$

The concept of the bending moment diagram is extremely important in structural analysis, for its use often gives a clear visual presentation of the load distribution within a frame structure.

As an example, a loaded beam and its accompanying shear force and bending moment diagrams are shown in fig. 1.5.

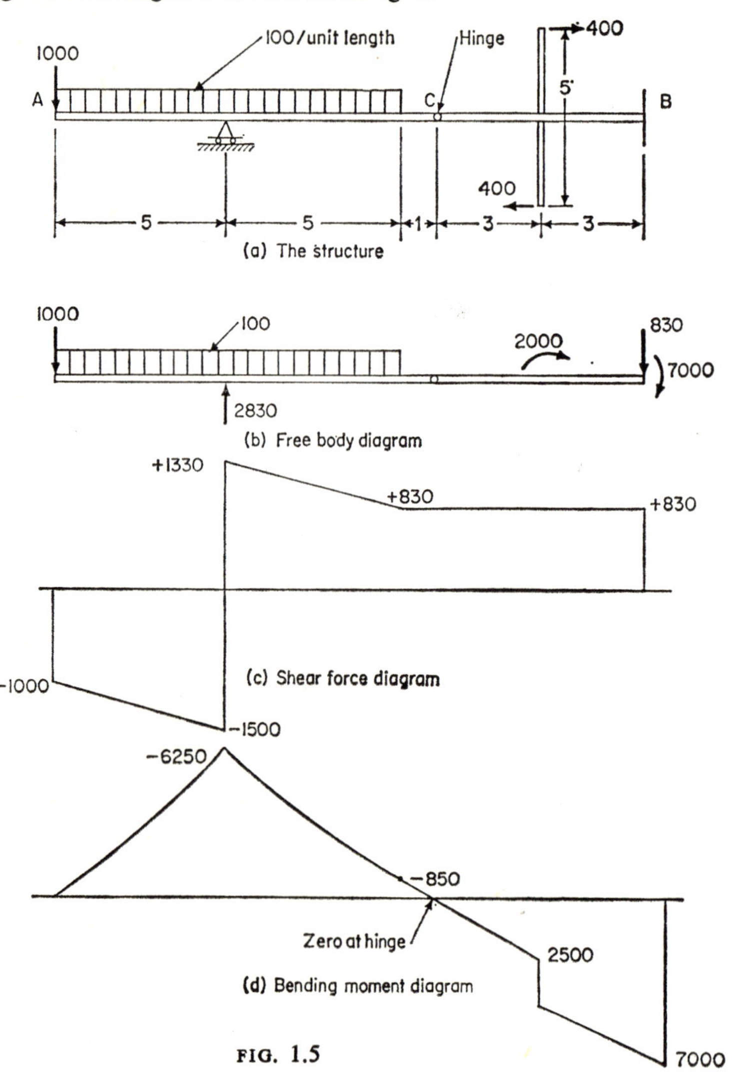

FIG. 1.5

1.2 Elasticity

1.2.1 Linear Elasticity

A load applied to an elastic structure will cause the structure to deform. If the structure is purely elastic, all the deformation will disappear when the load is removed. This does not imply that the deformation of the structure as the load was applied necessarily followed the straight line path A in fig. 1.6, for it could equally well have followed the non-linear paths B or C; it would still be an elastic structure as long as the paths taken during loading and unloading were the same.

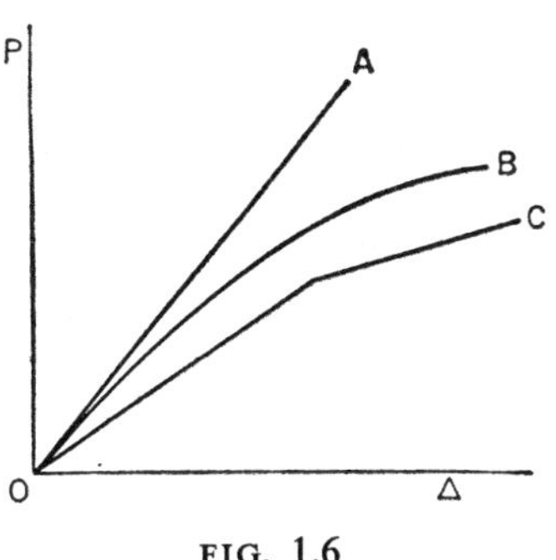

FIG. 1.6

A non-linear load-deflection path could result from one of two causes. Firstly, the material itself could have a non-linear stress–strain relationship: indeed there are very few engineering materials whose stress–strain relationship is linear over their entire range of behaviour. Typical stress–strain curves for mild steel and concrete are shown in fig. 1.7. However, if the maximum stress is kept low enough such materials do behave in an approximately linear manner. Non-linear behaviour may also occur in a structure even though it is constructed of a linear material, for it will occur in any structure whose deflections due to load are large enough to cause a redistribution of forces within itself. The behaviour of a suspension bridge is non-linear, and so is that of a guyed tower. Any structure with members that buckle or with tie wires that slacken would also behave non-linearly.

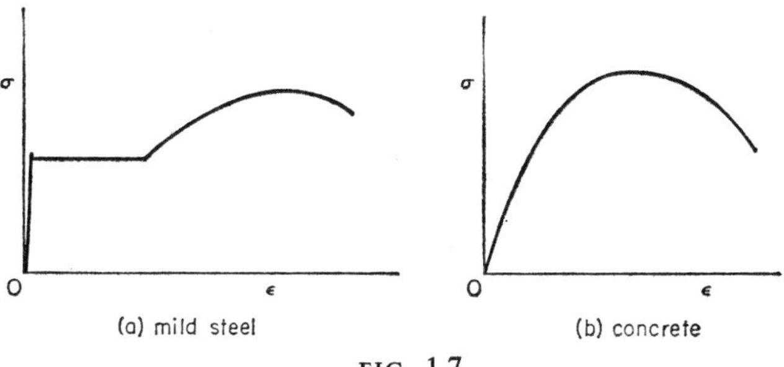

FIG. 1.7

The elastic theory of structural analysis presented in this book is restricted entirely to linear elastic structures, for by making this restriction the consequential simplification of the analysis is very great. The most important simplifications occur due to the principle of elastic superposition, which will be discussed in the next section.

1.2.2 Principles of Superposition

There are three principles of superposition concerned respectively with statics, with kinematics and with the linking of both in linearly elastic structures.

The *principle of superposition of forces* states that, for a statically determinate structure (that is, one for which the distribution of internal forces may be determined by considerations of statics alone) the reactions and internal forces due to various applied loads may be superposed. This principle is well known and familiar for it is fundamental to the use of statics, and follows directly from the equilibrium conditions given in equation (1.1), provided that any deflections of a structure caused by the forces applied to it are small and do not sensibly affect the distribution of forces within it. It says, for example, that if the reaction of the left-hand support of the beam shown in fig. 1.8 is

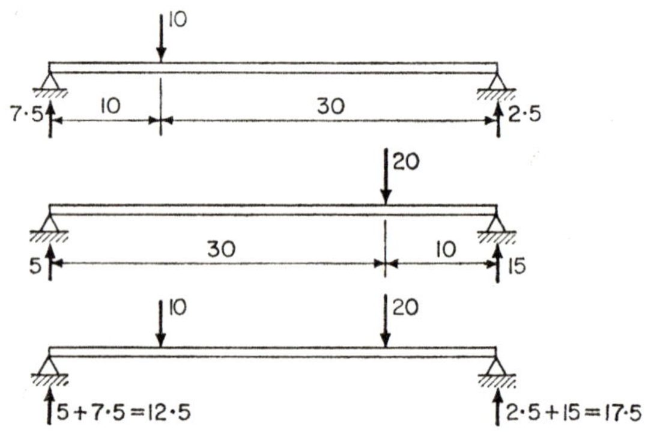

FIG. 1.8

7·5 with a load of 10, and 5 with a load of 20, then if both loads are applied simultaneously the reaction is $7{\cdot}5 + 5 = 12{\cdot}5$. In general, if forces P_1, $P_2 \ldots, P_n$ are applied to a statically determinate structure separately, and if each causes a reaction or internal force F at some point of $F_1 = a_1P_1$, $F_2 = a_2P_2$ and so on, where $a_1, a_2, \ldots, a_n$ are constants, then if all the forces are applied together the total value of F will be

$$\begin{aligned} F &= F_1 + F_2 + \cdots + F_n \\ &= a_1P_1 + a_2P_2 + \cdots + a_nP_n \end{aligned}$$

For a linearly elastic structure, force superposition is not confined to statically determinate structures but applies to indeterminate structures as well. This is a consequence of the principle of elastic superposition, to be discussed shortly.

The *principle of displacement superposition* states that if a series of small internal deformations is applied to a statically determinate structure then the displacement at some point due to all the internal deformations applied simultaneously is equal to the sum of the displacements at that point due to the deformations applied separately.† In other words, if small deformations $\delta_1, \delta_2, \ldots, \delta_n$ applied separately produce deflections $\Delta_1, \Delta_2, \ldots, \Delta_n$ at point A_1 where $\Delta_i = b_i\delta_i$ and where b_i is a constant, then the deflection when all the deformations are applied together is

$$\Delta = b_1\delta_1 + b_2\delta_2 + \cdots + b_n\delta_n$$

Note that the principle is an approximation which only applies if the displacements are small. As an example, consider the beam of fig. 1.9. Internal deformations δ_1 and δ_2 (which are in this case relative rotations) separately

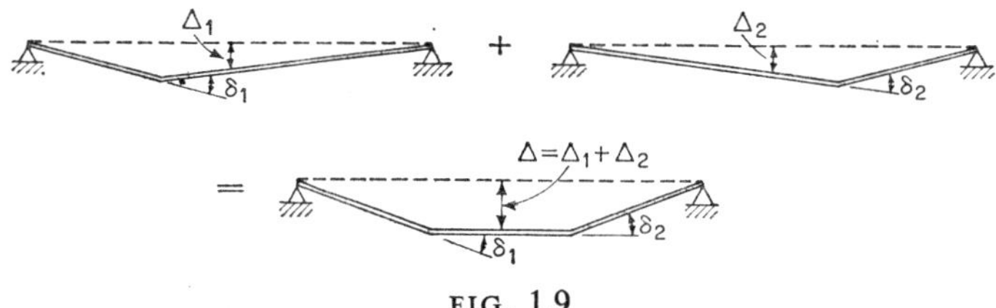

FIG. 1.9

produce central deflections of Δ_1 and Δ_2, while together the deflection is $\Delta = \Delta_1 + \Delta_2$. Fig. 1.10 shows an example in which the internal deformations occur not just at a point but are distributed over a portion of the structure. The total deformation of the end of the cantilever is due to the combination of Δ_1 due to the bending of the inner portion alone and Δ_2 due to the bending of the outer part.

The principles of force and displacement superposition were concerned solely with either the statics or the kinematics of a structure and not at all

† The restriction of the principle to statically determinate structures is not necessary but was made above in the interests of simplicity.

More generally, consider the deformable structure to be an n-degree of freedom mechanism, for which n independent sets of small displacements $\boldsymbol{\delta}_1, \boldsymbol{\delta}_2, \boldsymbol{\delta}_3, \ldots, \boldsymbol{\delta}_n$ may be found, where $\boldsymbol{\delta}_i$ is a vector (in the matrix sense) of dependent displacements. Then if each displacement set $\boldsymbol{\delta}_i$ causes a displacement Δ_i at some point A such that $\Delta_i = \mathbf{b}_i\boldsymbol{\delta}_i$ where $\mathbf{b}_i$ is a row vector of constants, then when any combination of sets of displacements $\boldsymbol{\delta}$ is applied, the total displacement at A is

$$\Delta = \mathbf{b}_1\boldsymbol{\delta}_1 + \mathbf{b}_2\boldsymbol{\delta}_2 + \cdots$$

with its elastic properties. If the load–displacement curve for a structure is linear, both statics and kinematics are linked in the *principle of elastic superposition* which states that for a linearly elastic structure, the displacements due to two force systems applied separately may be superposed when the force

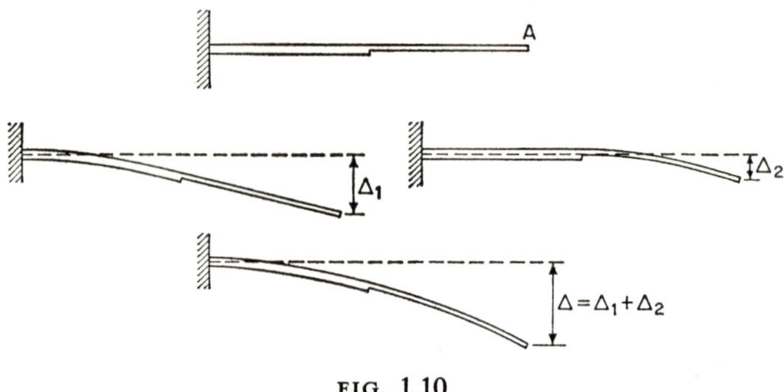

FIG. 1.10

systems are applied simultaneously. Thus if a number of forces $P_1 \ldots P_n$ are applied separately to a structure and if the deflection of a point A due to P_i is $\Delta_i = f_i P_i$ where f_i is a constant, then if all the forces are applied together the total deflection of A is

$$\Delta = f_1 P_1 + f_2 P_2 + \cdots + f_n P_n$$

This principle is illustrated by the cantilever of fig. 1.11. The ability to superimpose the effects of different loads and load magnitudes on a structure means that the structure can be analysed in general terms without having to consider specific load magnitudes until a late state of the analysis. More importantly perhaps, the load–deformation characteristics of each element of a complex structure can be analysed separately and independently in algebraic terms, with only the final stage of the analysis being concerned with the correlation of these element effects to give the properties of the structure as a whole.

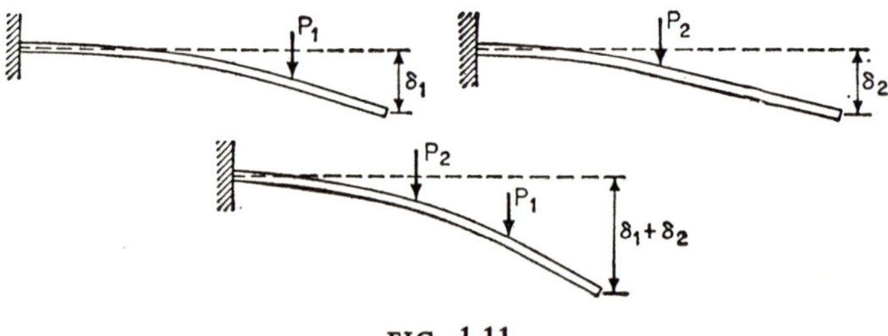

FIG. 1.11

1.2.3 Strain Energy

Consider a simple elastic system such as the cantilever of fig. 1.12(a), and suppose that a single load P, when applied gradually, gives a final deflection Δ. As the load is applied, the deflection will follow the path O–A of fig. 1.12(b). Suppose that before the load has built up to its final value it has an inter-

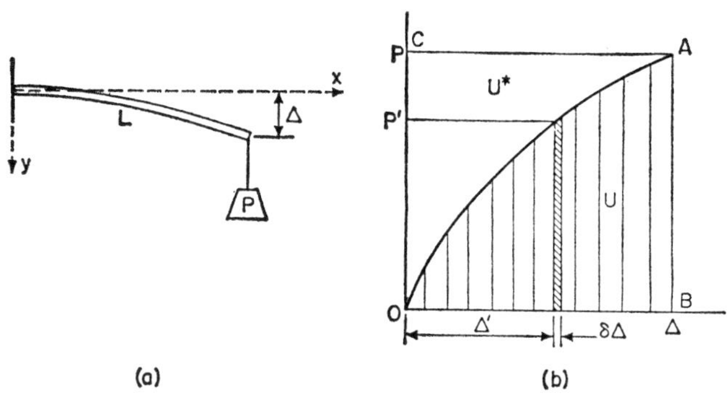

FIG. 1.12

mediate value P' with a corresponding deflection of Δ', and that a small increase in force δP at that stage causes an increase in deflection of $\delta\Delta$. The work done on the system due to this change will be

$$\delta W \approx P'\,\delta\Delta$$

The total work done in applying the load to the system is then

$$W = \int_0^{\Delta} P'(\Delta)\,\mathrm{d}\Delta$$

which is the area under the load–displacement curve. Evidently, if the load–displacement curve is a straight line, then

$$W = \tfrac{1}{2}P\Delta \tag{1.5}$$

For an elastic system the work is recoverable. It can be used, as in a diving board or the spring of a clock. Thus when the load is applied to the elastic system the work must be stored as an internal potential energy: this energy is called *strain energy*, and will be denoted by U. Hence

$$U = W$$

and the strain energy is the area OAB in fig. 1.12(b). Note that strain energy is always a positive quantity, no matter in what direction forces are applied.

The strain energy in a structure is also equal to the internal work done by the stresses throughout the structure acting through the corresponding strains. For a uniform prismatic bar of length L and cross-sectional area A subjected to an axial load P (with some intermediate value P') the stress will be P/A and the strain will be Δ/L, if Δ is the elongation of the bar. Then

$$U = \int_0^L \left(\int_0^\Delta \frac{P'}{A} \cdot \frac{\mathrm{d}\Delta}{L} \right) A \, \mathrm{d}x$$

If the bar is linearly elastic and has a Young's modulus E, then

$$U = \frac{1}{2} \cdot P \cdot \frac{\Delta}{L} = \frac{1}{2} \cdot \frac{P^2 L}{AE} \tag{1.6}$$

which is known as Clapeyron's theorem.

For a beam subject to bending, as in fig. 1.12(a), the strain energy $\bar{U}$ per unit length at any section is given by

$$\bar{U} = \int_0^K M' \, \mathrm{d}K$$

where M' is the moment and K is the curvature due to bending.

For a linearly elastic beam,

$$K = \frac{M'}{EI}$$

so

$$\bar{U} = \int_0^M \frac{M' \, \mathrm{d}M'}{EI} = \frac{1}{2} \cdot \frac{M^2}{EI}$$

and the strain energy for the whole beam is then

$$U = \frac{1}{2} \int_0^L \frac{M^2 \, \mathrm{d}x}{EI} \tag{1.7}$$

The area OAC of fig. 1.12(b) is called the *complementary energy* of the system and is denoted by U^*. It has no physical significance and so is not as useful a concept as that of strain energy. Nevertheless the use of complementary energy is fundamental to the general statement of certain important theorems in the theory of structural analysis, and it will be used for their derivation in Chapter 7. In any case, for a linear elastic system in which OA is a straight line, the strain energy is equal to the complementary energy, and

$$U^* = U$$

So far the discussion has been confined to an elastic system subjected to only one load. The concept of strain energy is, however, perfectly general and may be applied to structures subjected to any number of loads. For a linear elastic system, if all forces are applied to the structure simultaneously and gradually, the relationship between any one force and the displacement at a point may be represented by a diagram similar to fig. 1.12(b) except that OA is a straight line; and the principle of elastic superposition says that the final deflections will be the same no matter what the order of loading.

It is important to realise that the amounts of strain energy due to the application of individual forces cannot in general be superimposed. For example, suppose that in fig. 1.12(a) forces P_1 and P_2 are applied separately to the end of the cantilever. If the cantilever were linearly elastic, then the corresponding end deflections could be written

$$\Delta_1 = fP_1$$
$$\Delta_2 = fP_2$$

where f is a constant. The strain energy for each case would be

$$U_1 = \tfrac{1}{2}P_1\Delta_1 = \tfrac{1}{2}fP_1^2$$
$$U_2 = \tfrac{1}{2}fP_2^2$$

If, however, P_1 and P_2 are applied simultaneously, then the total strain energy would be

$$\begin{aligned} U &= \tfrac{1}{2}f(P_1 + P_2)^2 \\ &= U_1 + U_2 + fP_1P_2 \end{aligned}$$

which is not equal to the sum of the strain energy due to the separate application of the loads.

1.3 Deflection of Beams

1.3.1 Direct Integration

The theory of the elastic bending of beams depends on the following four assumptions:

(i) The internal and external forces of the beam are in equilibrium. This leads to the equation

$$\frac{d^2M}{dx^2} = -w(x) \tag{1.4}$$

(ii) The material properties of the beam are linear for the stress levels considered: the material obeys Hooke's law.

(iii) St. Venant's principle holds, that localised effects remain so. This is a reasonable assumption as long as the beam is not too short and deep.

(iv) The following kinematic conditions are satisfied:

(a) The deflections of the beam are sufficiently small that they do not affect the distribution of forces within it. This is generally true for most engineering structures, though not, for example, for axially loaded beams.

(b) The Bernoulli–Euler assumption holds, that initially plane sections remain plane. This leads directly to the well-known equation

$$\frac{M}{EI} = \frac{1}{R} = -\frac{\mathrm{d}^2 y}{\mathrm{d}x^2} \tag{1.8}$$

In fact, the presence of shear forces in the beam causes the cross-sections to distort, but this is not significant except for short, deep beams.

(c) Shear deformation is neglected. Once again this is a reasonable assumption except for short, deep beams such as spandrel beams beneath window openings, for example.

Equations (1.4) and (1.8) may be combined to give the fourth-order equation

$$\frac{\mathrm{d}^2}{\mathrm{d}x^2}\left(EI\frac{\mathrm{d}^2 y}{\mathrm{d}x^2}\right) = w(x) \tag{1.9}$$

However, starting directly from (1.8) we may write

$$\frac{\mathrm{d}y}{\mathrm{d}x} = -\int\frac{M}{EI}\,\mathrm{d}x + C_1 \tag{1.10}$$

$$y = -\iint\frac{M}{EI}\,\mathrm{d}x\,\mathrm{d}x + C_1 x + C_2 \tag{1.11}$$

where C_1 and C_2 are constants whose value may be obtained from the boundary conditions of the beam. The evaluation of the integral and the calculation of the constants give the beam deflection directly.

Direct integration is rarely used in practice as a means of obtaining the deflection of a beam unless its entire deflected shape is required. It is cumbersome to use and runs into difficulties where there is any discontinuity of loading or of section properties, though the former difficulty can be surmounted by the use of the Macaulay–Clebsch notation.

1.3.2 Moment Area

The method of moment area is essentially a different way of integrating equation (1.8); but the semigraphical interpretation of the integration process that is possible with the method gives it greater power and flexibility than the direct integration approach, particularly when the deflection or rotation of a single point on the beam is required.

If the slope of the beam at any point is θ, then equation (1.10) gives

$$\theta = -\int \frac{M}{EI}\,dx + C_1$$

From this we may say that

> The slope of a beam at some point relative to the slope at another is equal to the area of the M/EI diagram between the two points.

Strictly, the negative sign should be taken, but in practice the sign is not important as the direction of the slope is always obvious. Fig. 1.13 shows a cantilever for which the changes of slope are assumed to occur at finite points rather than continuously. It can be seen from the geometry of the system that

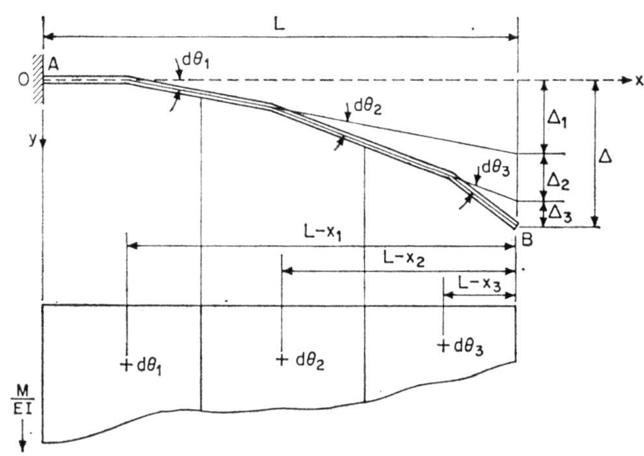

FIG. 1.13

the slope at B relative to A must be the sum of the intermediate slope changes $d\theta_1$, $d\theta_2$ and so on: this is an application of the principle of displacement superposition. It can similarly be seen from the diagram that the displacement of B relative to A is equal to the sum of the moments of the slope changes taken about B, i.e. that

$$\begin{aligned}\Delta &= \Delta_1 + \Delta_2 + \cdots \\ &= d\theta_1(L - x_1) + d\theta_2(L - x_2) + \cdots\end{aligned}$$

and in the limit

$$\Delta = \int (L - x)\frac{d\theta}{dx}\,dx = -\int \frac{M}{EI}(L - x)\,dx \qquad (1.12)$$

It is interesting to compare this equation with equation (1.11): the two expressions appear fundamentally different, but one may be obtained from the other by integrating by parts. In more visual terms we may now write that the deflection of a point **B** on a beam relative to the tangent at some other point A is equal to the moment of the *M/EI* diagram about B.

The following points should be noted about the method:

(i) It is the *M/EI* diagram, not the bending moment diagram, that is used, though for uniform beams the shapes of the two diagrams are the same.

(ii) There is no restriction on the method with regard to discontinuities of load or of beam cross section; but moment area cannot be carried across a kinematic discontinuity such as a hinge or sliding joint in a beam.

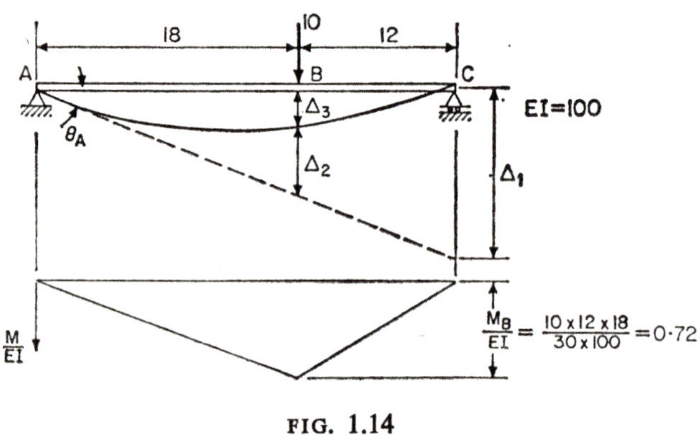

FIG. 1.14

(iii) The method does not give absolute slopes and deflections directly, but rather the relative slopes and deflections of one part of a beam relative to another. Corrections for rigid-body movements must be made to get absolute displacements.

(iv) Moment area is particularly good for spot values at discrete points. It is not particularly suitable for obtaining continuous distributions of displacements; but these are rarely required in practice.

(v) Apart from graphical solutions, the method is difficult to use where load or section property distributions have led to an *M/EI* diagram which is not a combination of simple geometric shapes.

To illustrate the method, consider the beam of fig. 1.14 for which it is required to find the deflection of the beam at B and its slope at A. The latter may be found directly as

$$\theta_A = \frac{\Delta_1}{30}$$

$$= \frac{1}{30} \times \text{moment of area of } M/EI \text{ diagram about C}$$

$$= \frac{1}{30} \times 0{\cdot}72 \left(\frac{18}{2} \times 18 + \frac{12}{2} \times 8\right) = 5{\cdot}04$$

By geometry the deflection at B must be

$$\Delta_3 = 18\theta_A - \Delta_2$$

$$\Delta_2 = \text{moment of left-hand part of } M/EI \text{ diagram about load point}$$

$$= 0{\cdot}72 \times \frac{18}{2} \times 6 = 38{\cdot}9$$

so that

$$\Delta_3 = 5{\cdot}04 \times 18 - 38{\cdot}9 = 51{\cdot}8$$

A frequently used variation of the moment-area method is the *conjugate-beam method* which uses the analogy between equations (1.3) and (1.8) as the basis of a formalised procedure which automatically takes into account the end slopes and deflections of a beam and hence derives absolute deflections instead of the relative deflections produced by the moment-area method. The beam is 'loaded' with the M/EI diagram, and the subsequent 'bending moments' are the deflections. The actual end conditions of the beam must, however, be replaced by a different set for the analogy between bending moment and deflection to hold: rotations become shears and deflections become moments.

There is little advantage in using the conjugate-beam approach rather than moment area, so that in practice an engineer will use whichever method is more familiar to him.

1.3.3 Superposition

The most straightforward way of calculating beam deflections is often to break down the beam or its loading into known simple cases and to superimpose the deflections caused for each. A list of simple cases is given in Table 1.1. The first two cases occur often enough to make it worth while remembering them; the others can easily be calculated in a few moments using, say,

TABLE 1.1 Standard cases for deflections of uniform beams

		End deflection Δ	End rotation θ
(a)		$\frac{PL^3}{3EI}$	$\frac{PL^2}{2EI}$
(b)		$\frac{ML^2}{2EI}$	$\frac{ML}{EI}$
(c)	w/unit length	$\frac{wL^4}{8EI}$	$\frac{wL^3}{6EI}$
		End rotation θ_a	End rotation θ_b
(d)		$\frac{Pab\,(L+b)}{6LEI}$	$\frac{Pab\,(L+a)}{6LEI}$
(e)		$\frac{ML}{3EI}$	$\frac{ML}{6EI}$
(f)	w/unit length	$\frac{wL^3}{24EI}$	$\frac{wL^3}{24EI}$

moment area. Two different superposition approaches may be used, based on either the principle of elastic superposition or the principle of displacement superposition.

An example of the use of elastic superposition is shown in fig. 1.15. The end deflection of the uniform cantilever is required due to loads P_1 and P_2. If P_1 is applied alone, the end deflection and slope are

$$\Delta_1 = \frac{P_1a^3}{3EI} + \frac{P_1a^2b}{2EI}; \qquad \theta_1 = \frac{P_1a^2}{2EI}$$

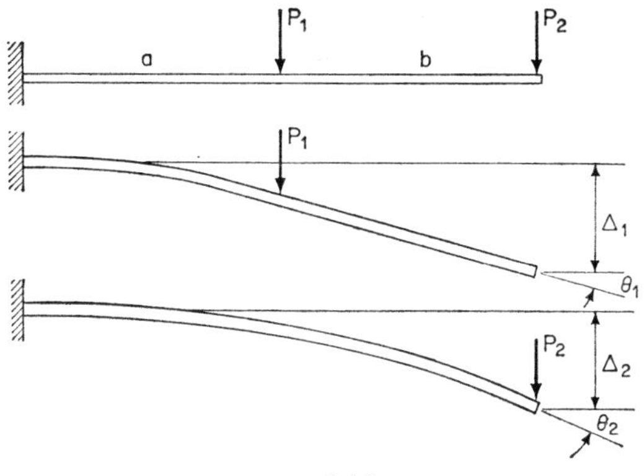

FIG. 1.15

If P_2 is applied, then the end deflection and slope become

$$\Delta_2 = \frac{P_2(a+b)^3}{3EI}; \qquad \theta_2 = \frac{P_2(a+b)^2}{2EI}$$

and the total deflection and slope when both loads are applied together are

$$\Delta = \Delta_1 + \Delta_2; \qquad \theta = \theta_1 + \theta_2$$

The use of the principle of displacement superposition is a more powerful approach to this type of problem, for a beam can be broken down into as many component elements as seems fit, and the displacement effects of the separate elements summed for the displacement of the whole. For example, the cantilever of fig. 1.16 may be split into two parts as shown. The shear force and moment acting on the end of the inner part AB are $P_1 + P_2$ and P_2b: the deflection and slope at B must then be, respectively

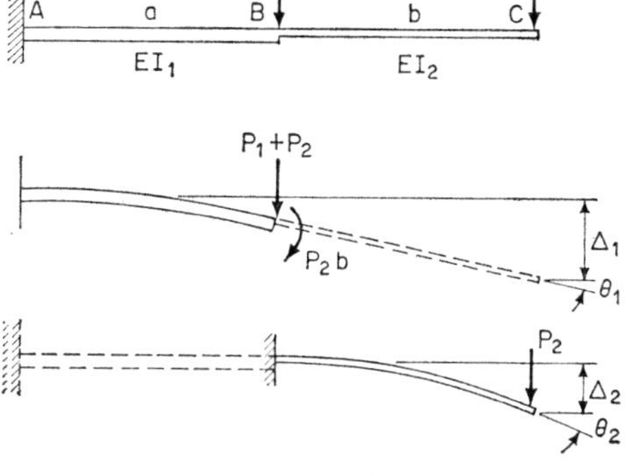

FIG. 1.16

$$\frac{(P_1 + P_2)a^3}{3EI_1} \quad \text{and} \quad \frac{(P_1 + P_2)a^2}{2EI_1}$$

If BC is taken to be rigid, then the deflection and slope at C due to the deformation of AB must be

$$\Delta_1 = \frac{(P_1 + P_2)a^3}{3EI} + \frac{(P_1 + P_2)a^2 b}{2EI}$$

and

$$\theta_1 = \frac{(P_1 + P_2)a^2}{2EI}$$

To this must be added the effect of the deformation of BC due to P_2. BC must be imagined to be built in at B: if this is done, the deflections at C will be

$$\Delta_2 = \frac{P_2 b^3}{3EI_2}$$

$$\theta_2 = \frac{P_2 b^2}{2EI_2}$$

Once again, this time by the principle of displacement superposition, the slope and deflection at C are equal to the sums of the component parts, or

$$\Delta = \Delta_1 + \Delta_2$$
$$\theta = \theta_1 + \theta_2$$

1.3.4 Numerical Integration

Although most beams occurring in civil engineering structures have a uniform cross-section, there are many instances in which non-uniform beams are used. A beam may be haunched, for example, it may have cover plates added to strengthen it in regions of high moment, or it may have to fulfil a design

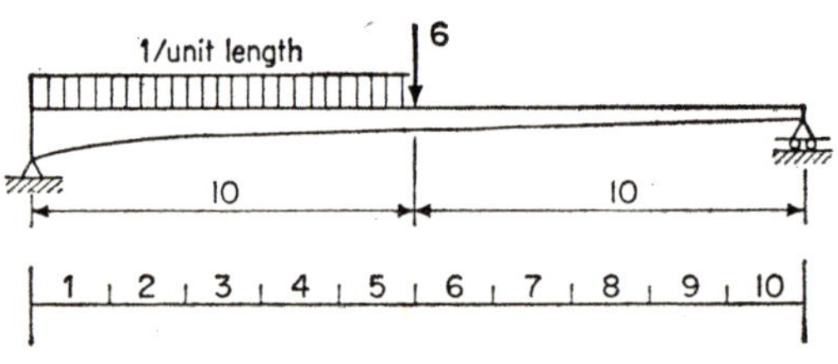

FIG. 1.17

requirement calling for an unusual shape. The deflection characteristics of some non-uniform beams may be found by one of the foregoing methods, or from sets of standard tables such as the PCA *Handbook of Frame Constants*; but if neither of these approaches can easily be used, then the problem must be solved by a numerical method.

Table 1.2 gives a numerical solution for the deflected shape of the beam shown in fig. 1.17, based on equation (1.11). If the deflection at only one

TABLE 1.2

1	2	3	4	5	6	7	8	9	10	11	12	13	14
Station No.	I	Load W	Shear $V=\sum W$	$M/\mathrm{d}x$ $=\sum V$	M Canti-lever moment	M Correc-tion moment	M Cor-rected moment	$\frac{M}{I}$	$\sum \frac{M}{I}$ $=\frac{\theta}{\mathrm{d}x}$	$\sum \frac{\theta}{\mathrm{d}x}$	Correc-tion	$\sum \frac{\theta}{\mathrm{d}x}$ Cor-rected	Beam deflec-tion $\times 10^6$
End						0					0		0
1	0·475	2		0	0	10·5	10·5	22·1		0	260	260	35
			2						22·1				
2	0·425	2		2	4	31·5	27·5	64·5		22·1	780	758	101
			4						86·6				
3	0·382	2		6	12	52·5	40·5	106·0		108·7	1300	1191	159
			6						192·6				
4	0·343	2		12	24	73·5	49·5	144·0		301·3	1820	1519	202
			8						336·6				
5	0·309	5		20	40	94·5	54·5	176·0		638	2340	1702	227
			13						512·6				
6	0·280	3		33	66	115·5	49·5	177·0		1151	2860	1709	228
			16						689·6				
7	0·255	0		49	98	136·5	38·5	151·0		1840	3380	1540	205
			16						840·6				
8	0·232	0		65	130	157·5	27·5	118·5		2681	3900	1219	162
			16						959·1				
9	0·216	0		81	162	178·5	16·5	76·4		3640	4420	780	104
			16						1035				
10	0·203	0		97	194	198·5	4·5	22·2		4675	4940	265	35
End			16	105	210	210·0			1057	5203	5200		0

point is required rather than the deflected shape of the whole beam, it would probably be easier and more accurate to base the numerical integration on equation (1.12) instead. Another approach would be to use the difference form of equation (1.8).

The beam must first be split up into a number of discrete elements. In this case, ten equal sections each 2 units long have been chosen. It may not always be convenient to choose equal element lengths, but wherever possible this should be done. The average moment of inertia for each section is now calculated and entered in column 2, while the load acting on the beam is divided into equivalent concentrated loads acting at the centres of the beam elements. These loads are set down in column 3, where it will be noticed that the concentrated load on the centre of the beam has been divided equally between adjacent sections.

The first step in the calculation is to find the bending moment distribution in the beam. In this example the bending moment at any point can easily be calculated directly, but for the sake of illustration it is found by numerical integration. The loads at each station are summed to find the shear (column 4) which is in turn summed and multiplied by the station interval to give a bending moment in column 6. This is not the actual bending moment in the beam, however, but the bending moment that would occur if the beam were cantilevered from its right-hand end. By extrapolation, the bending moment at the end of the beam is seen to be 210, whereas it should really be zero. Accordingly, a linear correction is made (column 7) varying from 210 to zero; and the difference between this and column 6 gives the true bending moment (column 8). A few check calculations will show that the figures in this column are very close to the true bending moment distribution.

In column 9, the bending moment is divided by the moment of inertia at each station. The same process of two integrations and a linear correction is then carried out, except that the interval width is omitted at each integration. Finally, column 13 is multiplied by 4 to allow for the omission of the interval widths and divided by the section stiffness $EI = 3 \times 10^7$. The final result in column 14 will have a maximum error of about 3%.

PROBLEMS

1.1 Use the method of virtual work to find the force in member AB of the truss shown in fig. 1.18. Suggested displacement system: let AB alone be shortened by a small amount, while all other members retain their original length.

1.2 Find the force in member 3–5 of the truss shown in fig. 1.19.

1.3 Draw shear force and bending moment diagrams for the beam shown in fig. 1.20, which has hinges at B and E.

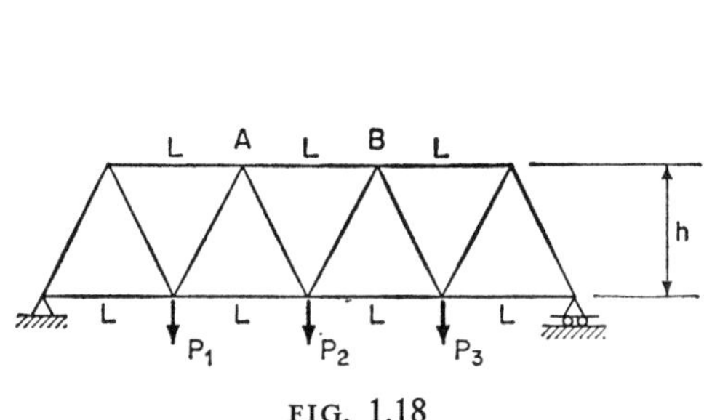

FIG. 1.18

FIG. 1.19

1.4 A construction bridge consists of a beam of length 50 simply supported at its ends. A Caterpillar tractor tows a compacting roller across the bridge: the tractor is a load of 1500 uniformly distributed over a length of 10 units, and the roller is a concentrated load of 1000 following a distance 10 behind the tractor. Find:

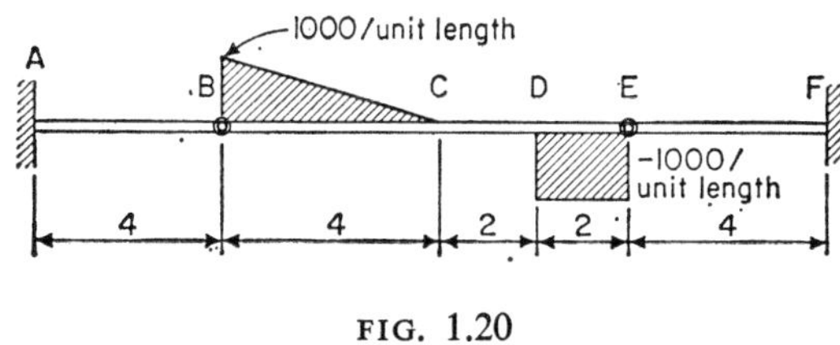

FIG. 1.20

(a) the maximum bending moment that can occur beneath the concentrated load, and the value of x at which this happens.
(b) the value of x for which the bending moment under the uniform load is a maximum, and the position and value of this maximum.

Answers: (a) $x = 30{\cdot}5$
$M = 21{,}000$
(b) $x = 24{\cdot}4$
$M = 18{,}840$
Position of maximum moment is 2·20 from left-hand edge of uniform load.

1.5 A uniform simply supported beam of length $(2a + 2b)$ is symmetrically loaded with two concentrated loads P at a distance a from either end. Find, both by the method of superposition, and by the method of moment area, the central deflection of the beam.

1.6 Direct integration of the equation $d^2y/dx^2 = -M/EI$ gives, for the deflection of the outer end B of a cantilever AB of length L,

$$\delta_B = -\iint_0^L \frac{M}{EI}\,dx\,dx \tag{1}$$

The moment area method gives

$$\delta_B = -\int_0^L \frac{M}{EI}(L - x)\,dx \tag{2}$$

Show the mathematical equivalence of statements (1) and (2).

1.7 A simply supported beam of length $2a$ carries a uniformly distributed load w over its entire length. Half the beam has a stiffness EI, while the other half has a stiffness $2EI$. Find the maximum deflection of the beam.
Answer: $0{\cdot}159wa^4/EI$

1.8 A long flexible uniform beam lies on a rigid horizontal surface. The weight/unit length of the beam is 100 and its section stiffness $EI = 10^7$. One end A of the beam is lifted by a crane and when it has risen a height 0·1, the beam fails in bending. Find how far the point of failure is from A.

Answer: 11·0

2

The Problem of Structural Analysis

2.1 Analysis and Design

2.1 Analysis and Design

Structural analysis is not design itself but is an integral part of it. The design of a structure is a process of synthesis which begins with the basic functional requirements of the structure, continues under the constraints of economic efficiency and structural feasibility, and concludes with the completed design. As soon as the initial ideas of a design have been formulated a rough analysis of various structural members must be made as a guide to their dimensions. Such calculations are sufficient until the design is almost complete: it is not until this stage that a thorough analysis of the entire structure is carried out to check that the overall strength and deflections of the structure lie within allowable limits. A rough, early analysis must necessarily be based on rules of thumb and on a series of gross approximations whose choice depends largely on experience and common sense, whereas for the final analysis of a structure much greater care must be taken in assessing the approximations used and in carrying out detailed and more rigorous calculations. Although this book seeks to some extent to give a feel for structural behaviour which is necessary for any approximate analysis, it is with the theory of more rigorous structural analysis that it is primarily concerned.

2.2 Structural Analysis

2.2 Structural Analysis

The aim of structural analysis is to seek to answer either or both of two fundamental questions about a structure. They are:

(a) Is it strong enough?
(b) Is it stiff enough?

The two questions can generally be treated separately, though there are instances, such as the dynamic analysis of structures subjected to earthquake forces, in which they are closely interrelated.

The first question asks how much load a structure can carry before failure. The problem is complicated by the fact that structures may fail in many different ways, quite apart from foundation failure. One structure might fail by rupture at some point, another by plastic yielding in its members and yet another by the instability of an overloaded column. In addition, the load to which a structure is actually subjected is often difficult to assess. Some loading is straightforward, such as the dead weight of a structure itself or the hydrostatic pressure applied to a dam, but other types of loading—wind loading for instance, or the load on a warehouse floor—are much more indeterminate.

The question of the stiffness of a structure may occur in various contexts. It may be that although a structure is strong enough its deflection under working load may be unacceptably large. The rails of an overhead crane may only be allowed to move relative to each other by a very small amount, for example, and the same is true of the sides of a lift well. Too great a flexibility in a space staircase would not please its users, too large a movement of the edge of a shell roof would cause difficulties in the design of the wall or window below it, and excessive deflections in a floor slab would cause an unacceptable level of cracking. In such cases a structure would have to be analysed for its deflection under working load conditions. The stiffness characteristics of a structure might also be needed in order to calculate its response to dynamic loading.

2.3 Working Stress and Ultimate Strength

The strength of a structure depends on the level of the resistance of its fabric to the stresses and internal forces caused by the loads applied to it. As the magnitude of the loads applied to a structure is increased, its behaviour will remain elastic (apart from local effects caused by built-in stresses due to welding and other factors) until at some point it begins to yield or a local fracture occurs. The structure will not generally fail completely at this stage, but the stress distribution within it will change, as will its stiffness. This redistribution of stress will continue with increasing load as further local failures due to yielding or fracture take place until eventually complete collapse occurs.

To determine the strength of a structure, two types of analysis can be used. The first, an *elastic analysis*, assumes that the stress distribution in the structure remains elastic at all times. The stress level is not allowed to rise above a certain arbitrarily set value called the *working stress*. As this approach does not take into account the stress distribution which occurs when a mild steel or reinforced concrete structure begins to yield, it often leads to a more conservative design than does the other approach, that of *ultimate strength analysis*. Such an analysis investigates the behaviour of the structure allowing for yielding in its members, and seeks to determine its ultimate strength. When the load to cause final collapse of the structure has been found, it is divided by a *load factor* to give an allowable load for the structure. The difference in concept between a load factor and a working stress (which may be some percentage of the yield stress), is important. In general it is preferable to use an ultimate analysis for fairly simple structures such as portal frames, beams and slabs as the calculations are simpler and a more economical design often results. However, ultimate strength analysis is not yet fully adequate for the design of more complicated structures such as multistorey frameworks or shell roofs. Certain difficulties arise in such cases—the problem of stability in frame structures, for instance—which tend to preclude its use so that an elastic analysis must be used instead. In any case, an elastic analysis is necessary to determine the deflections of a structure under working loads, should these be required.

Although most structures are analysed theoretically, there are instances in which a structure is too complicated or unusual for a theoretical analysis to be very reliable. In such circumstances it may be advisable to analyse the structure by constructing and testing a small scale model. There is not space in this book to give an adequate description of model analysis, but it is mentioned briefly as a useful method of analysis and one which every structural engineer should be prepared to use if necessary.

A model analysis can either be a direct test, in which case a plastic or reinforced mortar scale model is built and tested, or it may be an indirect test using a simple two-dimensional model to construct influence lines for the prototype structure. Direct tests need special laboratory facilities. They are slow and expensive, and are only justified economically for unusual and costly structures. Indirect tests are generally quick, easy to use and suited for use in an office, but they produce more limited information than direct tests. More detailed accounts of model analysis may be found in: Hendry, A. W., *Elements of Experimental Stress Analysis*, Pergamon, 1964; Charlton, T. M., *Model Analysis of Structures*, Spon, 1954.

2.4 Idealisation

It is impossible to analyse a structure in its entirety taking into account every detail. If the stiffening effects of partition walls, services, cupboards and so on were taken into account an analysis would be impossibly complicated quite apart from the fact that such effects are too indeterminate and unreliable to be readily analysed. Localised effects, too, such as the complicated stress distributions in a beam–column joint could not justifiably be incorporated in the overall analysis of a structure. Hence an actual structure as built is never analysed, but only an idealised conceptual model of the structure which is developed using various assumptions. Fig. 2.1(1) shows an actual structure, while (b) shows the idealised structure. First, the effect of all non-structural or non-load bearing parts is neglected. This is a conservative assumption, for the neglected parts are almost certain to add to the strength and stiffness of the structure. Next, the beams, slabs and columns of a frame structure are idealised into a two-dimensional pattern of negligible thickness, usually with no account being taken of end effects at joints (fig. 2.1(c)). For an elastic analysis it is also assumed that the structure behaves linearly, while equivalent assumptions may be used for an ultimate analysis.

The loads on a structure must also be idealised, which is probably the most indeterminate assumption of all. The dead loads on a structure due to its own weight can be determined quite accurately, but this is not true for the live loads applied later.

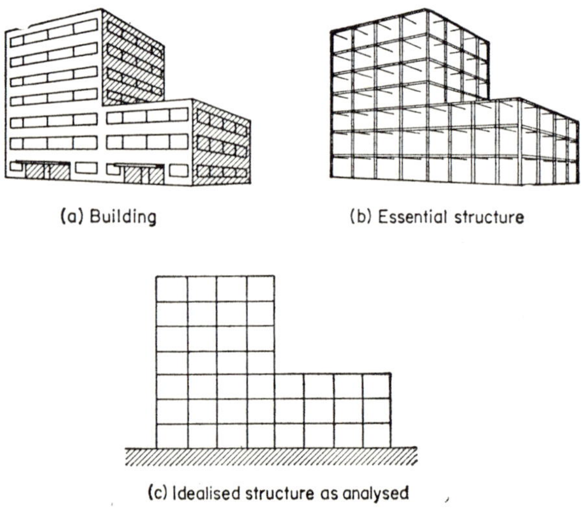

FIG. 2.1

The indeterminacy of the idealisation assumptions applied to a structure means that there is no point in analysing a structure to too great an accuracy. If the result is correct to within 10% this is usually sufficiently accurate, the exception to this being that if a detailed comparison of analyses is being made to determine, say, an optimum beam size, then an accuracy of about 2% would be needed.

2.5 Multiple-Element Structures

Although because of their nature the theoretical analysis of some structures such as dams or shell roofs must generally be carried out on the structure as a whole, many other structures naturally consist of a series of interlinked structural elements. Examples of multiple-element structures are shown in fig. 2.2.†

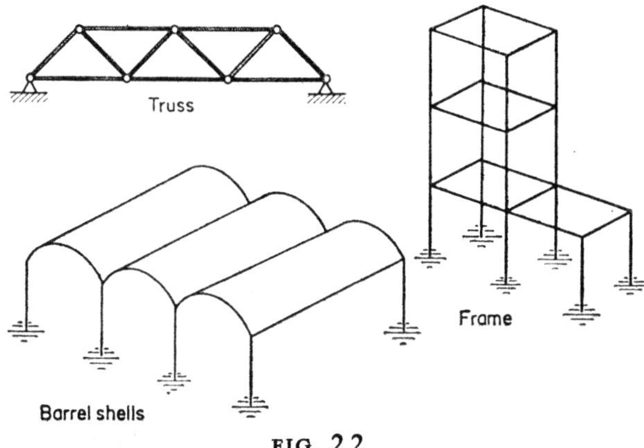

FIG. 2.2

Use of the principles of superposition allows the analysis of such structures to take place in two distinct steps:

(a) the analysis of individual members;
(b) the assembly of the effects of the separate members into an analysis of the whole.

Both the difficulties of calculation and the understanding of the behaviour of the structure are enormously simplified by this procedure. Each element with its own peculiarities of geometry can be dealt with separately and at leisure, and its structural behaviour can be expressed in terms of a few parameters. Only when the parameters for each element have been found are they assembled and the analysis of the total structure carried out. For example, the overbridge of fig. 2.3 would be analysed by first calculating the properties the individual members 1–2, 2–3, 2–4 and so on before combining them into an analysis of the whole.

This dual process of breaking the structure down into component parts, considering the behaviour of the parts and then synthesising their individual

† Even continuous structures may in fact be treated as multiple-element structures. A shell, for instance, may be idealised into a number of interconnected flat plate finite elements. Such a formulation would, however, need a computer for its solution.

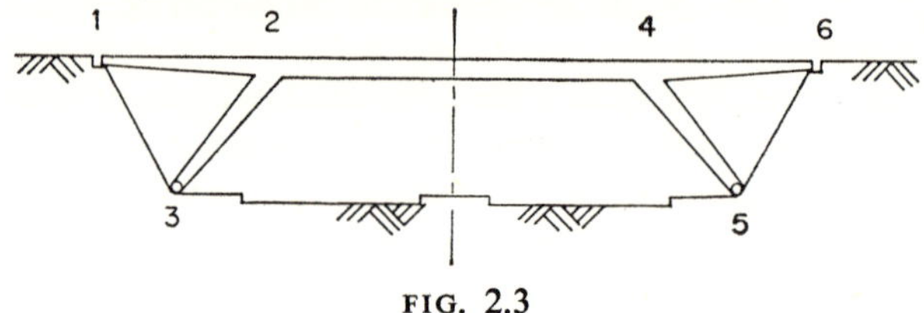

FIG. 2.3

effects to give the behaviour of the complete structure should be used whereever possible. It is a common and unnecessarily muddling mistake to try to understand and analyse too much at once.

2.6 Statically Indeterminate Structures

A structure is said to be *statically determinate* if all the internal forces and stresses caused by an applied load may be found solely from the equilibrium of the structure and the application of the principles of statics. If, on the other hand, the principles of statics alone are not sufficient to determine all

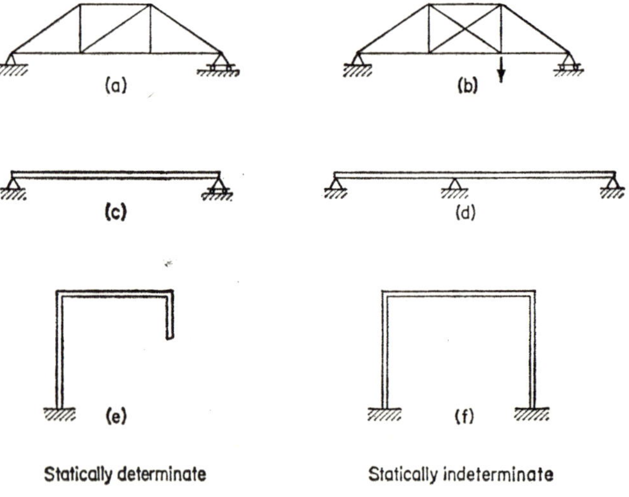

FIG. 2.4

the stresses in a structure, it is said to be *statically indeterminate.* Fig. 2.4 gives some examples of both types of structure. All internal forces in structures (a), (c) and (e) may be obtained from statics, no matter what the loading. The elastic properties of the structure are immaterial: the internal stresses would be the same whether it were linearly elastic or not, neither would they be affected if creep or plastic yield occurred. On the other hand, structures (d) and (f) are statically indeterminate for without a knowledge of their elastic behaviour their internal forces cannot be found. The pin-jointed truss (b) is also statically indeterminate as a whole, for even though the external reactions and some of the forces in the members caused by the applied force can be found, there are still certain forces which cannot be determined by statics alone.

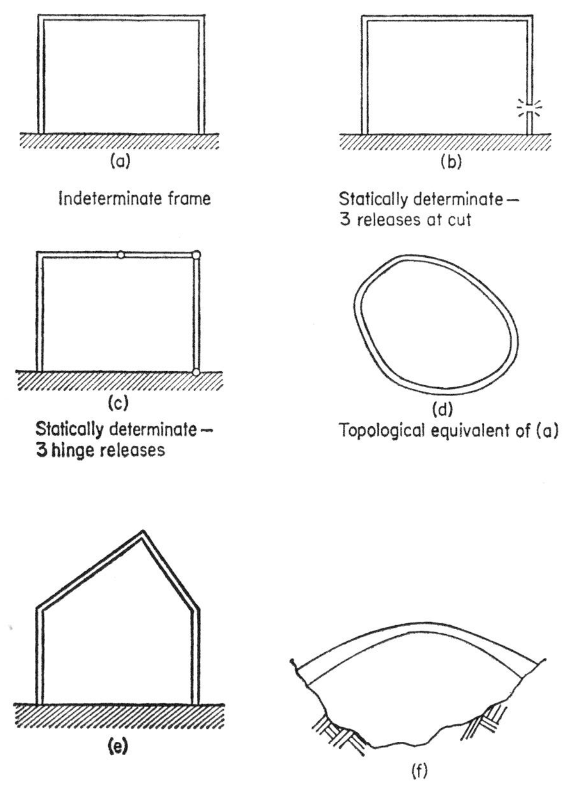

FIG. 2.5 Single ring structures

The *degree of indeterminacy* of a structure may be defined as the number of independent reactions and/or internal forces which must be known before the structure becomes statically determinate, or alternatively as the number of 'releases' which must be applied to the structure for it to become determinate. A 'release' is a hypothetical connection in a structure whose insertion means that some particular action cannot be carried across it. A hinge in a beam is a release as it transmits no moment, a cut in an axially loaded bar in a pin-jointed truss will transmit no force, and for a two-dimensional frame structure, a complete cut will provide three releases in a beam carrying axial load, shear and bending moment.

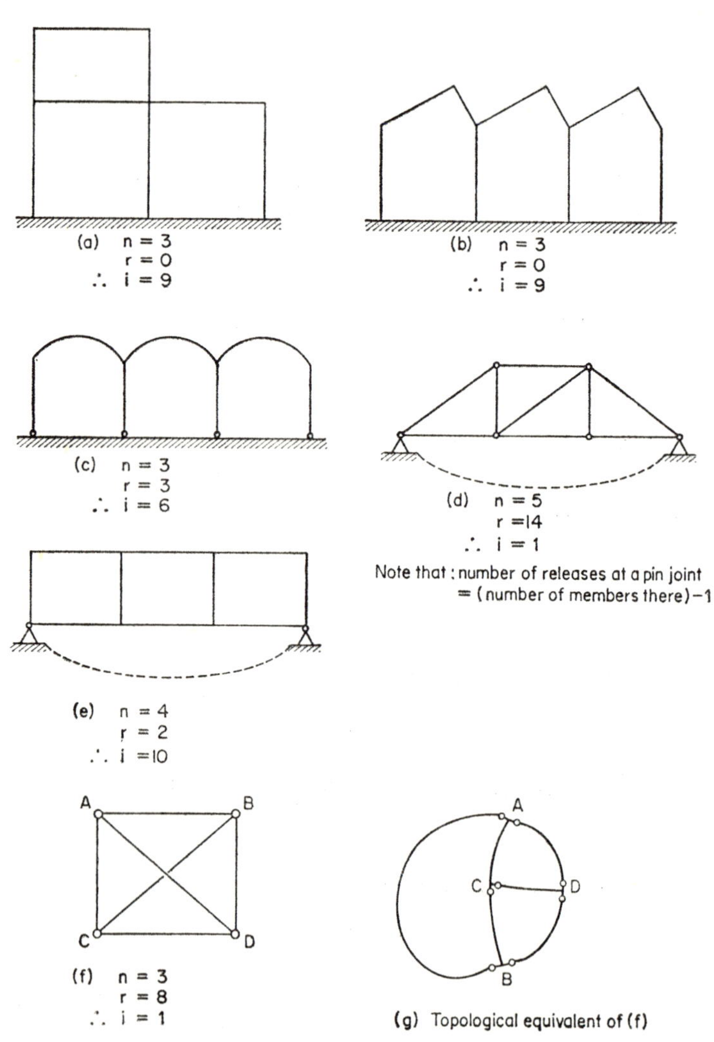

FIG. 2.6 Multiple ring structures

The portal frame of fig. 2.5 has both feet built in. A complete cut will make it statically determinate (fig. 2.5(b)) which therefore means (assuming loads are applied in its plane) that it is three times statically indeterminate. Any other three releases will also make it statically determinate: for instance, the three hinges shown in fig. 2.5(c). Indeed, as the frame and its foundation is topologically a rigid ring (fig. 2.5(d)) any stiff-jointed ring structure is three times indeterminate, and if a structure consists of a number of such rings joined together than the degree of indeterminacy of the whole is three times the number of rings. Examples of this are given in fig. 2.6; and the same figure shows some other two-dimensional structures whose rings are not completely stiff but which already contain some actual, not hypothetical, releases. If the number of rings is n and the number of actual releases is r then the degree of indeterminacy i of a two-dimensional structure is given by

$$i = 3n - r \tag{2.1}$$

A similar relation can be found for three-dimensional structures.†

2.7 Force and Displacement Methods

There are two basic approaches to elastic structural analysis: the *force method* and the *displacement method.* Both methods appear throughout the theory of structural analysis in so many different guises and bearing such a multiplicity of names that they are sometimes difficult to recognise, yet their basic separate characters remain. Although the two approaches are fundamentally different they are linked to one another by a tenuous complementary relationship which frequently helps the understanding of one or other method.

The two methods may be illustrated by the following example, The cantilever of fig. 2.7(a) is supported by a spring of stiffness k, and it is required to find the reaction R of the spring. The structure is evidently one times statically indeterminate.

The *force* approach begins by making the structure statically determinate. In this case the end reaction is assumed to be zero, and it is assumed initially that there is no connection between the spring and the cantilever. The end deflection Δ_1 due to the applied load w is now calculated by one of the methods outlined in the previous chapter (fig. 2.7(b)). Next, the applied load is removed and the reaction R is applied to the end of the cantilever and to the spring (fig. 2.7(c)): R is both the downwards force the cantilever applies to

† A fuller discussion of the problem of statical indeterminacy may be found elsewhere, for example in chapter 3 of Morice, P. B., *Linear Structural Analysis*, Thames & Hudson, 1959.

the spring and the upwards force the spring applies to the cantilever. The deflection Δ_2 and Δ_3 of the cantilever and the spring due to R are now calculated. The next step in the force method is to use the geometrical constraints of the system to provide equations from which the unknown forces can be found. In this case there is only one unknown, so that only one equation of

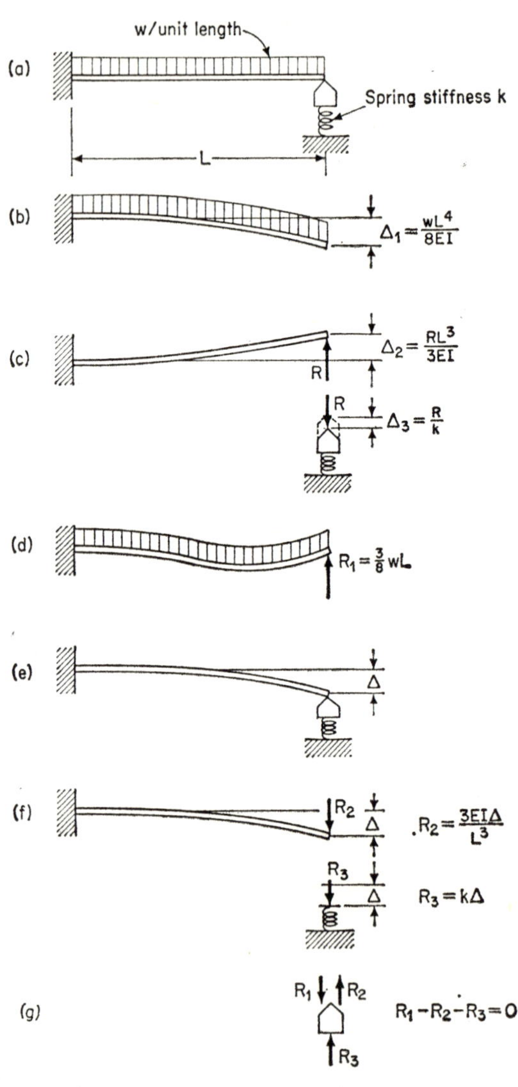

FIG. 2.7

geometry (or of kinematics) is needed. The required kinematic condition is that the downwards displacement of the spring must be the same as that of the cantilever. The total cantilever deflection is $\Delta_1 - \Delta_2$ and that of the spring is Δ_3, so that for geometric compatibility between the cantilever and the spring

$$\Delta_1 - \Delta_2 = \Delta_3$$

Writing the deflections in terms of forces,

$$\frac{wL^4}{8EI} - \frac{RL^3}{3EI} = \frac{R}{k}$$

whence the required result becomes

$$R = \frac{3kwL^4}{8(kL^3 + 3EI)} \tag{2.2}$$

The solution of the same problem by the *displacement* method begins by fixing the end of the cantilever against displacement and calculating the end reaction R_1 (fig. 2.7(d)). This is in itself a statically indeterminate problem which must be solved using the force method; but fortunately in this case the solution may be obtained directly from the previous result by substituting $k = \infty$ in equation (2.2). This requirement of the displacement method for an intermediate force-method solution is however only a slight disadvantage in practice as standard, known results would ordinarily be used at this stage. Next, the applied load w is removed and the connecting block between the spring and the end of the cantilever is given a downwards displacement Δ. This bends the cantilever, compresses the spring and implies the existence of forces R_2 and R_3 applied to them by the connecting block (fig. 2.7(e and f)). R_2 and R_3 are found in terms of Δ, and then an equilibrium equation is set up for the connecting block. The forces on it are shown in fig. 2.7(g), so that the equilibrium equation is

$$R_1 - R_2 - R_3 = 0$$

Rewriting this in terms of the unknown displacement Δ gives

$$\tfrac{3}{8}wL - \frac{3EI\Delta}{L} - k\Delta = 0$$

whence

$$\Delta = \frac{3wL}{8(3EI/L^3 + k)}$$

The required reaction is equal to R_3, so back-substitution into the expression for R_3 gives

$$R_3 = \frac{3kwL^4}{8(3EI + kL^3)}$$

as before.

Comparing the two solutions, the most obvious difference is that the force method chooses the force R as the primary unknown while the displacement method solves for the displacement Δ and then has to back-substitute to find the required reaction. The force method begins by inserting a release in the

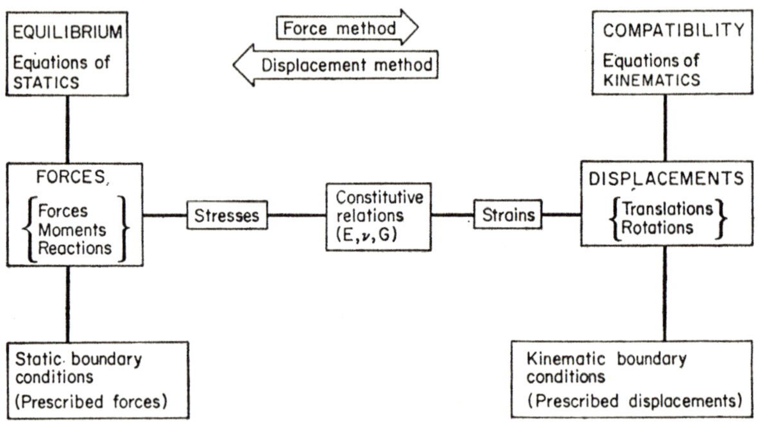

FIG. 2.8

structure to make it statically determinate. The resulting bending moment distribution fulfils the requirements of equilibrium, but the release has destroyed the geometric compatibility of the structure. This compatibility has to be restored by the application of a force across the release, whose magnitude is found by solving a compatibility equation written in terms of the unknown force. For a more complicated problem with more than one degree of indeterminacy the procedure is the same, except that there will now be as many unknown forces as the degree of indeterminacy, and that for each release there will be a separate equation of compatibility. These compatibility equations are written in terms of the unknown forces and solved simultaneously. But whereas the force method begins in general by making various forces zero, the displacement method starts its solution by making the necessary number of displacements zero: what that number is, will be discussed shortly. Compatibility, for the displacement method, is always maintained, but this time it is the equilibrium of the structure that is destroyed, to be restored later by the application of various unknown displacements. Equilibrium equations are now set up and written in terms of the unknown displacements: these equations are solved simultaneously for the displacements.

Every structure must fulfil the dual requirements of equilibrium and compatibility. The force method maintains the structure in equilibrium and uses the compatibility conditions for its solution, while the displacement method maintains the compatibility of the structure and obtains a solution by using the equilibrium conditions. The relationship between the two methods may be summed up by the diagram in fig. 2.8. Each method, as it were, links the

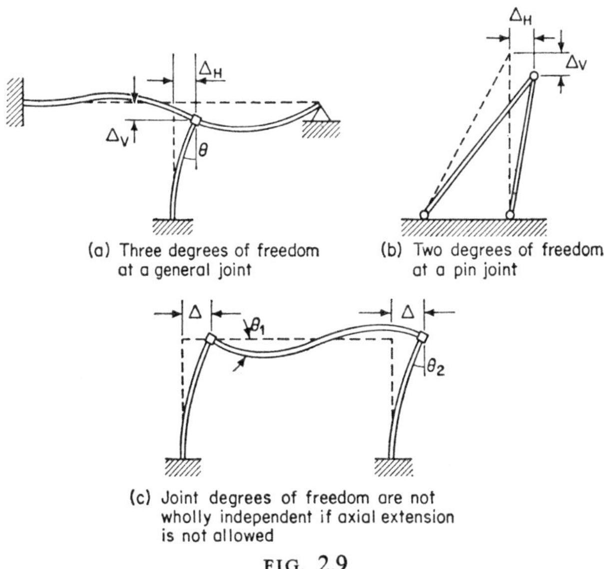

(a) Three degrees of freedom at a general joint

(b) Two degrees of freedom at a pin joint

(c) Joint degrees of freedom are not wholly independent if axial extension is not allowed

FIG. 2.9

two halves of the diagram: the force method ends up in the top right-hand corner with a set of kinematic equations in terms of forces, while the displacement method goes to the top left-hand corner with equilibrium equations written in terms of displacements rather than forces.

It has already been stated that the number of unknown forces that have to be found when using the force method is equal to the degree of indeterminacy of the structure. For the displacement method, the number of unknown displacements has nothing to do with the indeterminacy of the structure, but rather it is concerned with the number of degrees of freedom of its joints (or nodes). It is quite fortuitous that for the example of fig. 2.7

there happens to be the same number of unknown forces as of unknown displacements. A typical node in a two-dimensional fully elastic structure has three independent degrees of freedom: two linear displacements and one rotation (fig. 2.9(a)). For a pin joint between various members the rotation degree of freedom is trivial (the pin in the joint is perfectly free to rotate, and in doing so produces no moments) so such a joint has two degrees of freedom (fig. 2.9(b)). If, for a frame structure such as the portal frame of fig. 2.9(c), axial deformations of members are neglected, then some of the nodal degrees of freedom become dependent on one another and again the total number of degrees of freedom is reduced. The total number of independent nodal degrees of freedom in a structure may be called the degree of kinematic indeterminacy of the structure. This is the number of displacement unknowns that have to be found and the number of simultaneous equations that must be solved.

The various steps in the force and displacement methods may now be analysed and compared in the following scheme:

	FORCE METHOD	DISPLACEMENT METHOD
1.	Assess the degree of static indeterminacy of the structure.	Assess the degree of kinematic indeterminacy of the structure.
2.	Choose the required number of unknown forces and set them equal to zero. This makes the structure statically determinate.	Choose the required number of unknown displacements and set them equal to zero. This makes the structure kinematically determinate at the nodes.
3.	Calculate the displacements of the statically determinate structure due to the applied loads, and hence find the errors in compatibility.	Calculate the forces at the nodes in the kinematically determinate structure due to the applied loads, and hence find the errors in equilibrium.
4.	Apply the unknown forces one at a time and find the displacements due to each.	Apply the unknown displacements one at a time and find the nodal forces due to each.
5.	Set up equations of compatibility in terms of the unknown forces and solve them. The force distribution for the entire structure can then be found by application of statics.	Set up equations of equilibrium in in terms of the unknown displacements and solve them. The nodal displacements of the entire structure will now be known.
6.	Back-substitute for displacements if required.	Back-substitute for forces.

Some structures are more easily analysed by the force method and some by the displacement method. The pin-jointed truss of fig. 2.10(a) has one redundant member and so the force method would produce only one equation of compatibility. The displacement method, on the other hand, would need to solve the vertical and horizontal displacements at joints 2, 3, 4 and 5 and horizontal displacement at joint 6; nine unknowns in all which would require nine simultaneous equations. The structure of fig. 2.10(b), however, has a degree of indeterminacy of four but only two unknown displacements—the vertical and horizontal movements of the joint. It would obviously be quicker

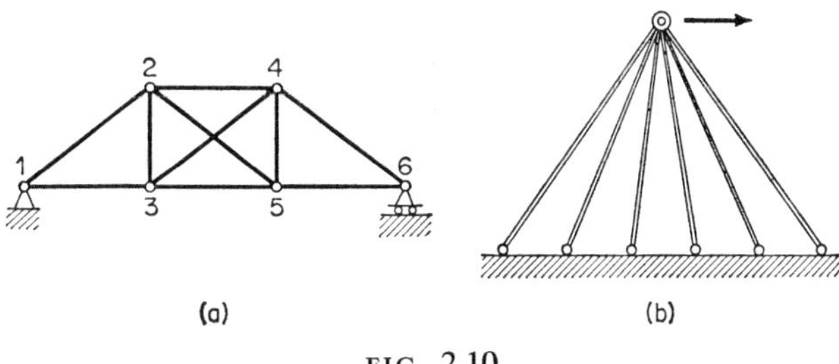

FIG. 2.10

to analyse this structure by the displacement method even though a back-substitution step would be necessary to produce internal forces from the displacements. In practice, the variants of the displacement method are more often used than those of the force method as they are easier to formulate in a systematic way.

It should also be noted that in some unusual problems mixed methods have been used, solving for both forces and displacements as unknowns. Such instances are, however, very rare.

PROBLEMS

2.1 Using the method of moment area or otherwise, find the bending moments at the ends and the centre of a uniform beam of length L, rigidly fixed at both ends,

(a) for a central concentrated load P,
(b) for a uniformly distributed load w.

2.2 A beam of length $4L$ has both ends built-in and is subjected to a central concentrated load P. The central half of the beam has a sectional stiffness EI, while cover plates on the outer quarters raise the stiffness there to $2EI$. Find the deflection of the centre of the beam.

2.3 A two-span continuous beam of length $2L$ is simply supported at A, B and C and has a section stiffness EI. Span BC has a length L and is subjected to a vertical concentrated load P at its centre. Find the bending moment and the rotation at B:

(a) by the force method, with the reaction at A as the unknown;
(b) by the method of displacements, with the rotation at B as the unknown;
(c) by the force method, with the banding moment at B as the unknown.

2.4 For the beam described in Problem 2.3, take $L = 10$, $P = 10$, $EI = 10^9$. If the simple support beneath B is elastic with a stiffness of $K = 50$ (where stiffness is defined as force to cause unit displacement), find the reaction and deflection at B.

3

Beams

3.1 The Functions of a Beam

In this chapter, a beam is regarded not so much as a horizontal member spanning a certain distance and subjected to various loads and end conditions, but rather, as one element of a multi-element frame structure. Neither, in this context, need the word 'beam' be reserved solely for horizontal members, for it can be used equally well for inclined or vertical members or indeed for any member in a frame structure with an approximately linear moment–rotation relationship.

As a member of a structure, a beam has one or both of two possible functions. They are, as shown in fig. 3.1:

(a) to transmit forces and moments from one part of the structure to another;
(b) to carry a load and to transmit its effects into the rest of the structure.

If the structure as a whole is statically indeterminate, both functions require the calculation of force-deformation relations for the beam, though the form of these relations depends on whether the force or the displacement method is to be used for the analysis of the structure. In general, the force method will require the flexibility characteristics of the beam; that is, its displacements due to unit applied forces; while the displacement method will need its stiffness characteristics, or the forces produced by the application of unit displacements. The flexibility and stiffness characteristics of a beam are of course not independent of one another but are different expressions of its basic properties.

3.2 Force Method Requirements

It will be remembered from the previous chapter that the basic approach to the force method of analysis is generally to begin by making a structure statically determinate by releasing the conditions of compatibility at various points. The compatibility errors for the determinate structure are then removed by the application of redundant forces to the structure. The beam-element information required for a force-method analysis will therefore be of a statically determinate nature, and the end deflections of a beam will need to be known in terms of forces so that the compatibility equations can be written.

There are two simple statically determinate beam elements: the cantilever and the simply supported beam. For a cantilever element, the information required for the two beam functions is the deflection and rotation of the end

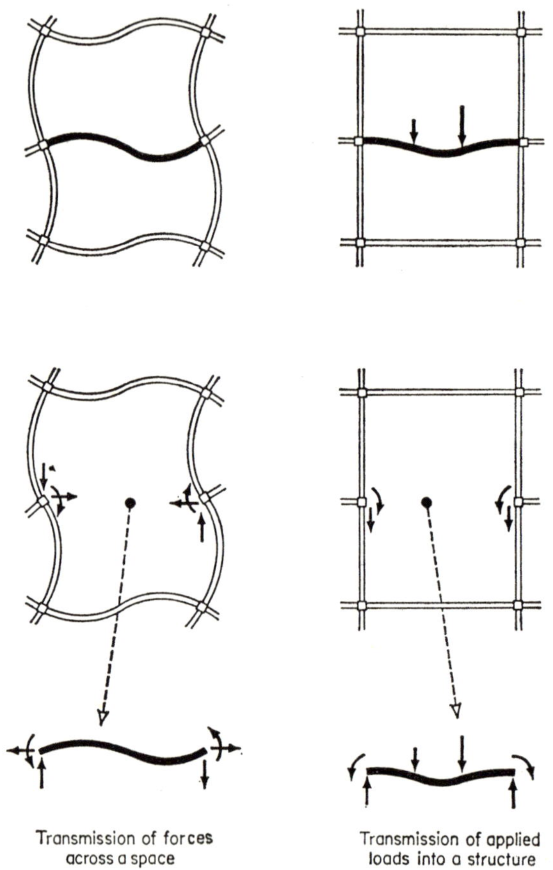

FIG. 3.1 The functions of a beam

of the cantilever due firstly to any loads applied along the length of the cantilever and secondly to a unit transverse force and a unit moment applied at its end (fig. 3.2(a)). For a simply supported element the end rotations are required due to transverse applied loads and due to unit end moments (fig.

3.2(b)). The calculations involved in finding this information are quite simple: any of the methods of Section 1.3 can be used successfully, though whenever possible standard results from Table 1.1 should be used.

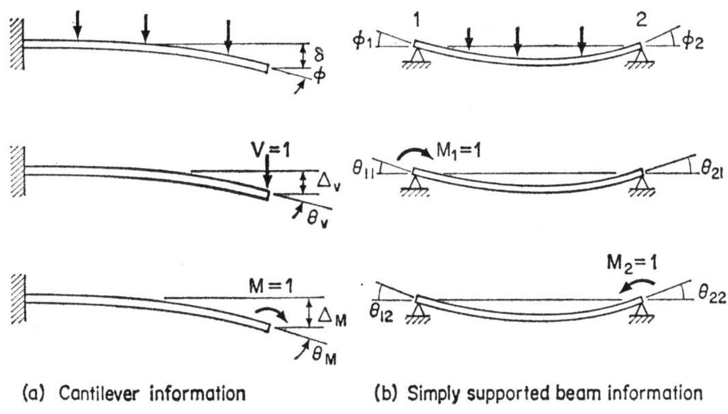

FIG. 3.2 Basic information required for the Force Method

3.3 Displacement Method Requirements—Stiffness and Carry-Over Factors

The unknowns for the displacement analysis of a frame structure are generally the rotational and linear displacements of the nodes (joints) of the structure. The analysis begins by holding these displacements to zero when the applied load is put on the structure and calculating the resulting end moments of the beam elements. It then proceeds by removing the applied load, applying a displacement—at this stage unknown and so expressed algebraically—to each node in turn and calculating the beam-end moments that result. These moments expressed in terms of the nodal displacements are then used to set up compatibility equations from which the unknown displacements are found. Thus the items of information needed for the two functions of a beam element in a displacement-solution analysis are the end moments on a beam due to applied loads with both ends of the beam fixed, and the end moments due to independent linear and rotational displacements of the beam ends (fig. 3.3). This latter type of information will be dealt with first, with the applied load effects being left to the next section.

Whatever the actual end displacements of a beam element, they can be represented by suitable combinations of the four displacements shown in fig. 3.3(b–e), except for displacements along the beam axis, which are not

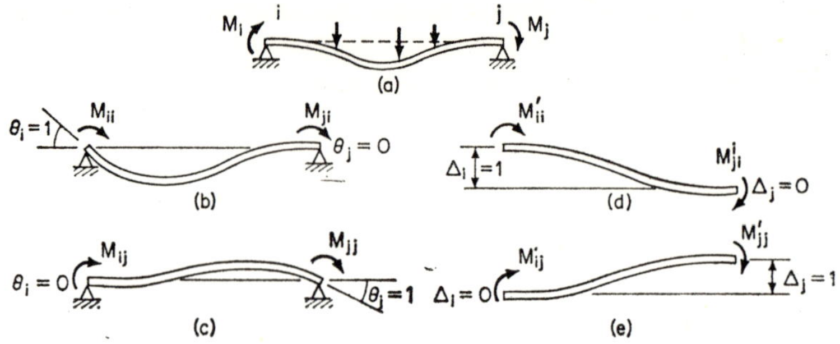

FIG. 3.3 Basic information required for the Displacement Method

considered here. However, there are only two, not four, basic displaced shapes of the beam; for, as shown in fig. 3.4, any transverse displacement Δ can be represented by two end rotations φ, where

$$\varphi = \frac{\Delta}{L}$$

The angle φ is known as the *chord rotation*: it is a useful concept as by allowing a linear displacement to be represented by a rotation in this manner the basic moment–displacement relations for the beam can be expressed in a particularly compact form.

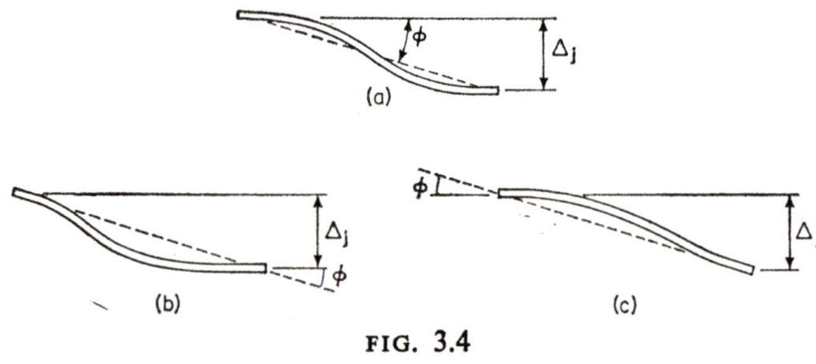

FIG. 3.4

For a displacement-method frame analysis it is important to adhere rigidly to a sign convention for rotations and moments. The convention used in this book is:

(a) All joint rotations and chord rotations are positive in a clockwise direction.

(b) All moments acting on the ends of the beam elements are positive clockwise.

It is important that this sign convention for beam end moments is not confused with the notation used in Chapter 1 for the internal bending moments in a beam, which were taken to be positive if the beam sagged. The moments considered here are specific values imagined to be applied externally to the ends of the beam, whereas the convention of Chapter 1 refers to bending moments which are continuous functions of internal forces within the beam.

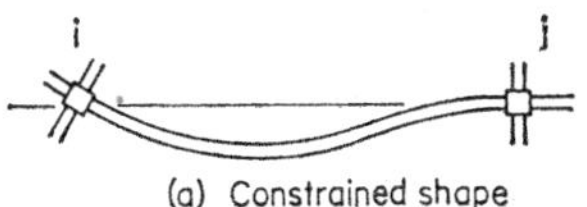

(a) Constrained shape

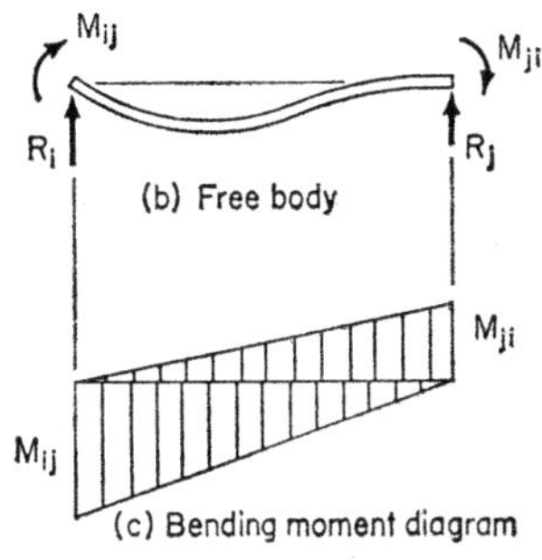

(b) Free body

(c) Bending moment diagram

FIG. 3.5

Now consider the beam shown in fig. 3.5, connecting joints i and j in a structure. If joint i is rotated through an angle θ_i and joint j is not allowed to rotate, the beam will be constrained into the deflected shape of fig. 3.5(a), and it will be acted on by the forces shown in (b). If the beam is elastic, there will be a linear relationship between θ_i and the beam-end moment at i, which may be written

$$M_{ij} = S_{ij}\theta_i \tag{3.1}$$

The constant S_{ij} is known as a *stiffness factor* for the beam, and it may be defined as the moment at i due to unit rotation at i. The effect of the application of the moment M_{ij} also carries across to the other end of the beam and induces a moment M_{ji} there. M_{ji} must evidently be proportional to M_{ij}, and this can be expressed by writing

$$\begin{aligned} M_{ji} &= C_{ij}M_{ij} \\ &= C_{ij}S_{ij}\theta_i \end{aligned} \tag{3.2}$$

The constant C_{ij} is called a *carry-over factor* as it expresses the carry-over of effects from i to j. A beam will have two stiffness factors, S_{ij} and S_{ji} corresponding to rotations of joints i and j, and two carry-over factors. However, only three of the four factors are independent for they are connected by the relationship

$$C_{ij}S_{ij} = C_{ji}S_{ji} \tag{3.3}$$

for reasons which will be explained in the next chapter.

For a uniform beam with section stiffness EI and length L, the stiffness and carry-over factors may readily be found using the moment–area method. If joint i is rotated while j is not, as in fig. 3.5, then

(a) the slope at i relative to j is θ_i, and
(b) the deflection of i relative to j is 0.

From these two conditions, and using the bending moment diagram of fig. 3.5(c) in which the effects of M_{ij} and M_{ji} are shown separately, the following two equations can be written:

$$\frac{M_{ij}}{EI}\cdot\frac{L}{2} - \frac{M_{ji}}{EI}\cdot\frac{L}{2} = \theta_i$$

$$\frac{M_{ij}L}{2EI}\cdot\frac{L}{3} - \frac{M_{ji}L}{2EI}\cdot\frac{2L}{3} = 0$$

Whence

$$M_{ij} = \frac{4EI}{L}\theta_i$$

and

$$M_{ji} = \tfrac{1}{2}M_{ij}$$

Hence in this case

$$S_{ij} = \frac{4EI}{L} \qquad (3.4)$$

$$C_{ij} = \tfrac{1}{2}$$

and as the beam is symmetrical we also have, for rotation of j and no rotation of i,

$$S_{ji} = \frac{4EI}{L}$$

$$C_{ji} = \tfrac{1}{2}$$

These results are often used, and should be remembered.

Beams which are non-uniform will, of course, have different stiffness and carry-over factors. Any of the methods given in Chapter 1 may be used for calculating the factors, the choice depending as usual on preference and the nature of the beam. Superposition or one of the analytic methods should be used where possible, but for beams with a more complex variation of section properties numerical integration can always be used to find stiffness and carry-over factors. It should be noted, however, that tables for the stiffness and carry-over factors for haunched beams may be found in various publications, the best source being the *PCA Handbook of Frame Constants.*

To illustrate the calculation procedure, the stiffness and carry-over factors will now be found for the beam shown in fig. 3.6 using the moment–area method, although the method of superposition could equally well have been used. First, j is fixed and i is given a unit rotation. The relative rotation between i and j in unity so that the total area of the M/EI diagram is zero. Hence

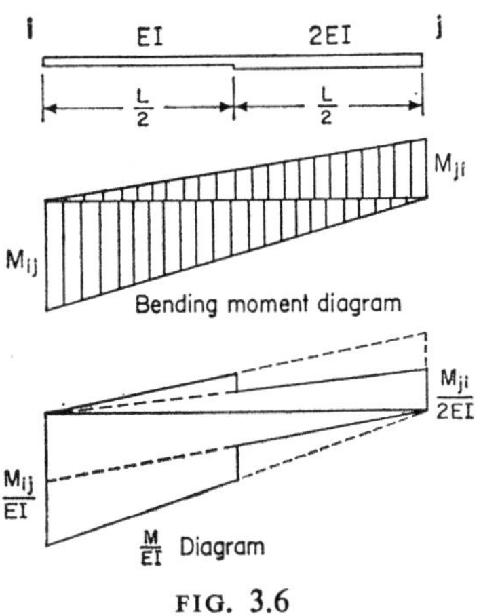

FIG. 3.6

$$\frac{M'_{ij}}{EI}\cdot\frac{L}{2}-\frac{M'_{ij}}{4EI}\cdot\frac{L}{4}-\frac{M'_{ji}}{2EI}\cdot\frac{L}{2}-\frac{M'_{ji}}{4EI}\cdot\frac{L}{4}=1 \tag{3.5}$$

The primes distinguish these moments from those obtained later by rotating j. The deflection of i relative to the tangent at j is zero, so taking moments of the M/EI diagram about i,

$$\frac{M'_{ij}L}{2EI}\cdot\frac{L}{3}-\frac{M'_{ij}L}{16EI}\cdot\frac{2L}{3}-\frac{M'_{ji}L}{4EI}\cdot\frac{2L}{3}-\frac{M'_{ji}L}{16EI}\cdot\frac{L}{3}=0 \tag{3.6}$$

This equation gives

$$M'_{ji}=\tfrac{2}{3}M'_{ij}$$

or

$$C_{ij}=\tfrac{2}{3}$$

and substituting this into (3.5) gives

$$S_{ij}=M'_{ij}=\frac{48EI}{11L}=4{\cdot}364\,\frac{EI}{L}$$

If now i is fixed and j is given a unit rotation, the bending moment and M/EI diagrams will have the same form as before although the values of the end moments will no longer be the same: let them be denoted by M''_{ij} and M''_{ji}.

Equation (3.5) will still apply except that it must be written in terms of the new moments, but a new deflection condition appears, that the deflection of j relative to i is zero. Taking the moment of the M/EI diagram about j, we obtain

$$\frac{M''_{ij}L}{2EI}\cdot\frac{2L}{3}-\frac{M''_{ij}L}{16EI}\cdot\frac{L}{3}-\frac{M''_{ji}L}{4EI}\cdot\frac{L}{3}-\frac{M''_{ji}L}{16EI}\cdot\frac{2L}{3}=0$$

From this

$$M''_{ij}=\tfrac{2}{5}M_{ji}$$

so

$$C_{ji}=\tfrac{2}{5}$$

Substituting this in (3.5) gives

$$S_{ji}=M''_{ji}=\frac{80EI}{11L}=7{\cdot}273\frac{EI}{L}$$

As a check on the solution, we can verify that (3.3) holds. It is, of course, advisable in all structural analysis calculations to check results wherever it is possible to do so.

3.4 Non-Rigid End Connections

Beam members of a structure may sometimes not be rigidly fixed to the nodes or foundations at which they terminate but may be connected to them by a hinge or some other flexible connection. In such a case the stiffness and

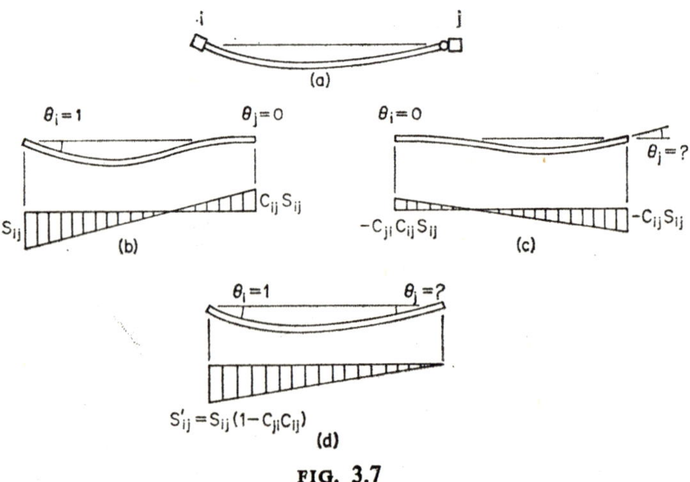

FIG. 3.7

carry-over factors for the rigidly connected beam must be modified to take into account the actual end conditions.

The expression for the modified stiffness factor for a beam pinned at one end is particularly simple. Such a beam is shown in fig. 3.7(a), where it is fixed at i and pinned at j. The modified stiffness factor S'_{ij} is, of course, the end moment at i when the beam is given a unit rotation at i. If the beam were fixed at j the moments would be as in fig. 3.7(b); so to bring the moment at j to zero a rotation must be applied at j sufficient to produce an equal and opposite moment at j (fig. 3.7(c)). The two deflection and moment diagrams are now superimposed in fig. 3.7(d) to give the required result, which is that

$$S'_{ij} = S_{ij}(1 - C_{ji}C_{ji}) \tag{3.7}$$

As there is no moment at j, the modified carry-over factor C'_{ij} must be zero; and S'_{ji} must also be zero as a rotation of node j (fig. 3.7(a)) produces no moment in the beam. For a uniform beam for which $S_{ij} = 4EI/L$ and the carry-over factors are a half,

$$S'_{ij} = \frac{4EI}{L}\left(1 - \frac{1}{2} \times \frac{1}{2}\right) = \frac{3}{4}\cdot\frac{4EI}{L} = \frac{3EI}{L} \tag{3.8}$$

This value is worth remembering.

If the beam-joint connection at j allows the beam end to displace but not rotate, as in fig. 3.8, an expression similar to (3.7) may be obtained fairly easily; but as the general expression is rather too complicated to be useful it will not be derived here. The beam is statically determinate, however, and has a constant bending moment which has the same distribution as that for a cantilever with a moment applied to its end. Hence if the beam is uniform we may use Table 1.1(b) to write

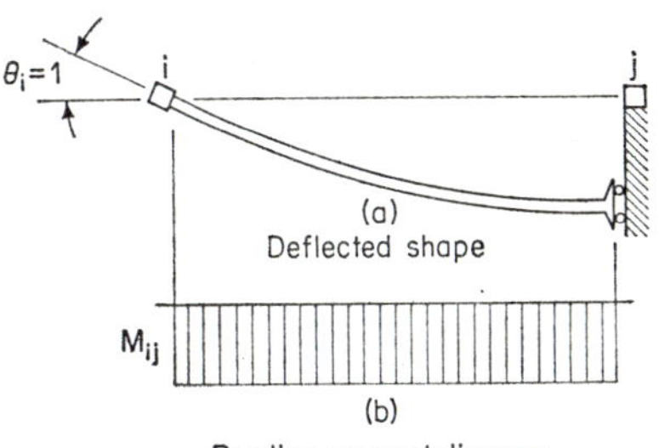

FIG. 3.8

$$\theta_i = 1 = \frac{M_{ij}L}{EI}$$

and hence

$$S'_{ij} = \frac{EI}{L} \tag{3.9}$$

From the bending moment diagram we can see that the carry-over factor must be

$$C'_{ij} = -1 \tag{3.10}$$

It sometimes happens that it is convenient to analyse a symmetrically loaded structure taking displacements on either side of the centre-line of the structure not as independent unknowns but as displacements equal and opposite to one another. If this is so, the ends of any beam whose centre-line coincides

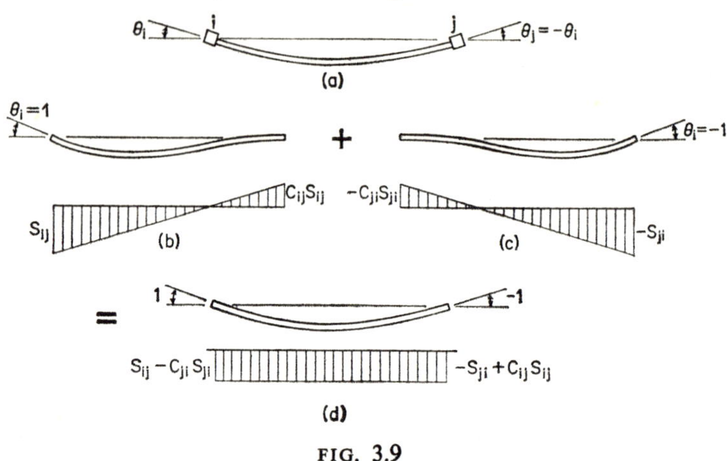

FIG. 3.9

with the centre-line of the structure must be subjected to equal and opposite displacements at its ends. Such a beam is shown in fig. 3.9: the beam itself must of course be symmetric. Superimposing cases (b) and (c) in the figure to get (d), it can be seen that the modified stiffness factor must be

$$S'_{ij} = S_{ij} - C_{ji}S_{ji}$$

which may be written because of symmetry or equation (3.3)

$$S'_{ij} = S_{ij}(1 - C_{ij}) \tag{3.11}$$

For a uniform beam this expression becomes

$$S'_{ij} = \frac{4EI}{L}\left(1 - \frac{1}{2}\right) = \frac{2EI}{L} \tag{3.12}$$

The bending moment diagram of fig. 3.9(d) shows that the modified carry-over factor must be

$$C'_{ij} = -1$$

though in this case the meaning of the carry-over factor must be treated with caution.

For uniform beams, the results found in this section are summarised in Table 3.1.

TABLE 3.1 Modified stiffness and carry-over factors for uniform beams

	End conditions	S'_{ij}	S'_{ji}	S'_{ij}	S'_{ji}
(a)	i j	$\frac{4EI}{L}$	$\frac{4EI}{L}$	$\frac{1}{2}$	$\frac{1}{2}$
(b)	i j	$\frac{3EI}{L}$	0	0	–
(c)	i j	$\frac{2EI}{L}$	$\frac{2EI}{L}$	−1	−1
(d)	i j	$\frac{EI}{L}$	$\frac{EI}{L}$	−1	−1
(e)	i j	0	0	—	—

3.5 Displacement Method Requirements—Fixed-End Moments

We now come to the second requirement of a beam element, that of sustaining applied loads along its length and transmitting these effects into the rest of the structure. For the beam shown in fig. 3.10, the actual load on the structure (a) could be replaced by a statically equivalent set of node moments and forces (b) such that the total behaviour of the structure would be precisely the same. Fig. 3.10(c) shows the beam treated as a free body: because it *is* a free body the system of forces in (c) is self-consistent and self-equilibrating, so that if this set of forces were subtracted from those in fig. 3.10(a) the result would be

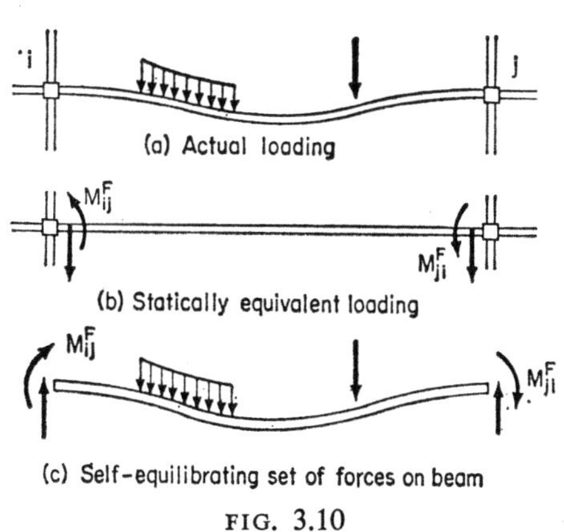

(a) Actual loading
(b) Statically equivalent loading
(c) Self-equilibrating set of forces on beam

FIG. 3.10

the distribution in (b), with no other forces elsewhere in the structure being affected. The moments M_{ij}^F and M_{ji}^F occurring on either end of the beam in fig. 3.10(c) are called *fixed-end moments*: they are the beam-end moments occurring at i and j due to some transverse applied load when joints i and j are fixed against rotation.

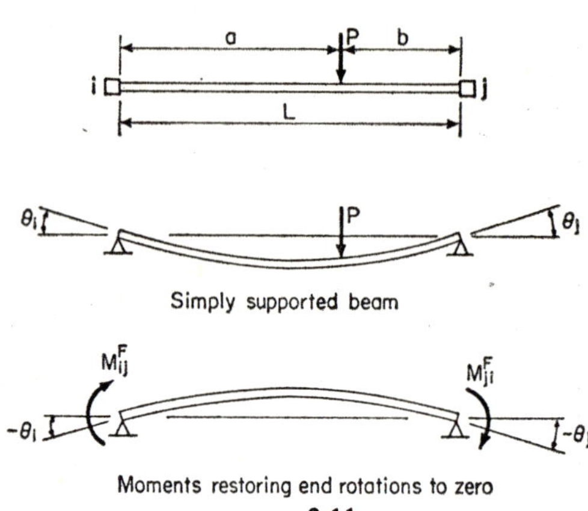

FIG. 3.11

For a uniform beam the fixed-end moments due to any loading can be found if the fixed-end moments due to a concentrated load at some arbitrary point on the span are known (fig. 3.11). We know from Table 1.1(d) that if the beam were simply supported, the end rotations would be

$$\theta_i = \frac{Pab(L+b)}{6LEI}$$

$$\theta_j = -\frac{Pab(L+a)}{6LEI}$$

using our convention that clockwise relations are positive. These end rotations are now restored to zero by the application of fixed-end moments M_{ij}^F and M_{ji}^F to the beam. Thus

$$M_{ij}^F = -(S_{ij}\theta_i + C_{ji}S_{ji}\theta_j) \tag{3.13}$$

$$= -\left(\frac{4EI}{L}(L+b) - \frac{2EI}{L}(L+a)\right)\frac{Pab}{6LEI}$$

$$= -\frac{Pab^2}{L^2} \tag{3.14}$$

Similarly

$$M_{ji}^F = +\frac{Pa^2b}{L^2}$$

From these results two particular cases follow. If the concentrated force is at the centre of the beam (fig. 3.12), $a = b = L/2$ and

$$M_{ij}^F = -\frac{PL}{8}$$
$$M_{ij}^F = +\frac{PL}{8} \tag{3.15}$$

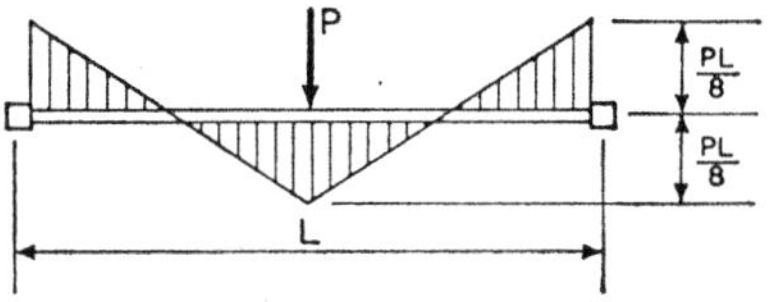

FIG. 3.12

For a uniformly distributed load w, elements of force $w\,dx$ are integrated to give

$$\begin{aligned} M_{ij}^F &= -\int_0^L \frac{wx(L-x)^2}{L^2}\,dx = -\frac{w}{L^2}\left(\frac{x^2}{2}L^2 - \tfrac{2}{3}x^3L + \frac{x^4}{4}\right)\Bigg|_0^L \\ &= -\frac{wL^2}{12} \end{aligned} \tag{3.16}$$

Similarly

$$M_{ji}^F = +\frac{wL^2}{12}$$

For non-uniform beams the best approach is still to find the simply supported end rotations for a beam and then to apply equation (3.13), for the stiffness and carry-over factors of the beam must in any case be known or calculated before a displacement-type analysis of the structure can be carried out. Consider, for example, the beam of fig. 3.14(a) which will be solved by superposition. First, assume the beam is fixed at its centre but free at its ends (fig. 3.14(b)). Then from Table 1.1(a) and (c), the end deflections and rotations are

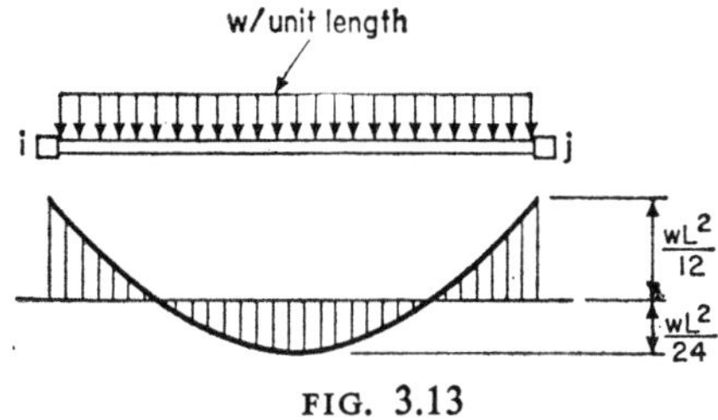

FIG. 3.13

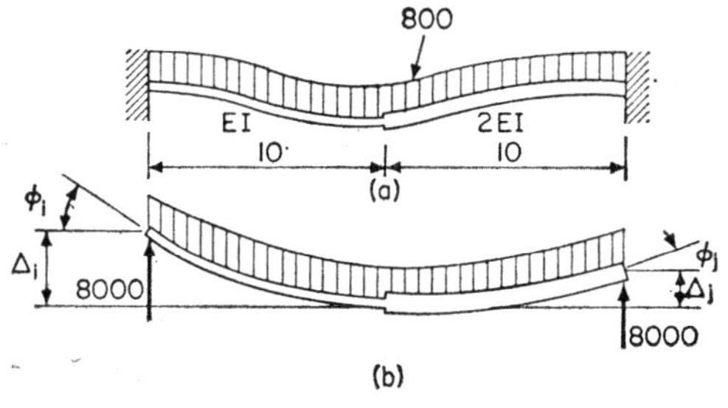

FIG. 3.14

$$\varphi_i = \frac{8000 \times 10^2}{2EI} - \frac{800 \times 10^3}{6EI} = \frac{8 \times 10^5}{3EI}$$

$$\varphi_j = -\frac{4 \times 10^5}{3EI}$$

$$\Delta_i = \frac{8000 \times 10^3}{3EI} - \frac{800 \times 10^4}{8EI} = \frac{5}{3} \times \frac{10^6}{EI}$$

$$\Delta_j = \frac{5}{6} \times \frac{10^6}{EI}$$

Now the actual end rotations θ_i and θ_j which would occur if the beam were simply supported at its ends are, from the geometry of the problem,

$$\theta_i = \varphi_i - \frac{\Delta_i - \Delta_j}{20} = \frac{8 \times 10^5}{3EI} - \frac{5 \times 10^5}{12EI} = \frac{2{\cdot}25 \times 10^5}{EI}$$

$$\theta_j = \varphi_j - \frac{\Delta_i - \Delta_j}{20} = -\frac{4 \times 10^5}{3EI} - \frac{5 \times 10^5}{12EI} = -\frac{1{\cdot}75 \times 10^5}{EI}$$

We already know from the example on page 55 that

$$S_{ij} = 4{\cdot}364 \frac{EI}{20} = 0{\cdot}218EI$$

$$S_{ji} = 7{\cdot}273 \frac{EI}{20} = 0{\cdot}364EI$$

$$C_{ij} = \frac{2}{3}$$

$$C_{ji} = \frac{2}{5}$$

Hence using the equation (3.13),

$$M^F_{ij} = -0{\cdot}218EI \times \frac{2{\cdot}25 \times 10^5}{EI} + \frac{2}{5} \times 0{\cdot}364EI \times \frac{1{\cdot}75 \times 10^5}{EI}$$

$$= -23{,}700$$

$$M^F_{ji} = 0{\cdot}364EI \times \frac{1{\cdot}75 \times 10^5}{EI} - \frac{2}{3} \times 0{\cdot}218EI \times \frac{2{\cdot}25 \times 10^5}{EI}$$

$$= +30{,}800$$

These figures should be compared with those for a uniform beam: in such a case the fixed-end moments are $\pm 26{,}700$ which fall, as they should, between the two results for the non-uniform beam of the present example.

If the fixed-end moments are known for a beam element fully fixed at its ends, it is a simple matter to modify them to take into account a pin joint at one end of the beam. Suppose, for instance, the fixed-end moment $M_{ij}^{F'}$ is required for the beam of fig. 3.15(a), and the fixed-end moments M_{ij}^F and M_{ji}^F are known if the beam is fixed at i and j. To make the end moment

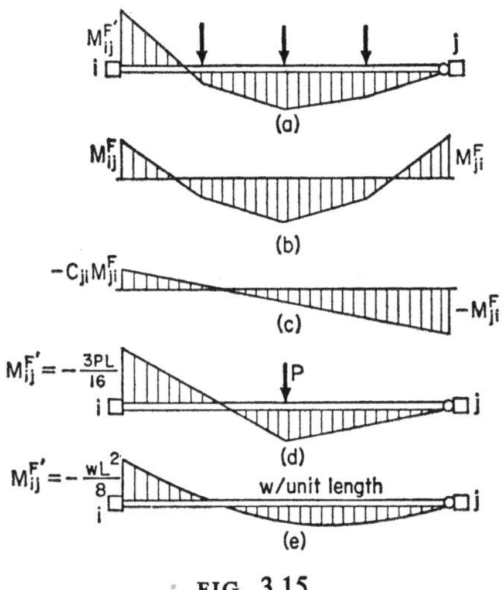

FIG. 3.15

at j equal to zero, a moment of $-M_{ji}^F$ must be superimposed (fig. 3.15(c)): this produces a moment at i of $-C_{ji}M_{ji}^F$ so that

$$M_{ij}^{F'} = M_{ij}^F - C_{ji}M_{ji}^F \tag{3.17}$$

for the pin-ended beam. For a uniform beam pinned at j with a concentrated central load P (fig. 3.15(d)) the modified fixed-end moment will be

$$M_{ij}^{F'} = -\frac{PL}{8} - \frac{1}{2}\cdot\frac{PL}{8} = -\frac{3PL}{16}$$

and for a uniform beam pinned at j subjected to a uniformly distributed load w,

$$M_{ij}^{F'} = -\frac{wL^2}{12} - \frac{1}{2}\cdot\frac{wL^2}{12} = -\frac{wL^2}{8}$$

If the pin were at the other end of the beam the modified fixed-end moments would of course have the same values but different signs.

PROBLEMS

3.1 A non-uniform beam AB of length 20 is cantilevered from A. If a unit vertical force is applied at B, the rotation and vertical deflection at B are 0·0186 and 0·211 while if a unit moment is applied at B the rotation and vertical deflection are 0·00192 and 0·0186.

If the same beam is now simply supported at A and B, find the slopes of the beam at A and B due to a unit moment applied at A, and also the slopes at the ends due to a unit moment at B.

Answer: (a) 0·000527, 0·000493
(b) 0·000403, 0·000587

3.2 A uniform beam of length 40 for which $EI = 100$ is strengthened at both ends by cover plates welded on to the outer quarter spans. The cover plates increase EI to 180. Find the stiffness and carry-over factors of the beam.

Answer: 15·6, 0·586

3.3 Points B and C lie at the third points of a beam AD of length 30, which is part of a structure. AB and CD have a section stiffness EI of 100 while a cover plate is added between C and D, bringing its stiffness to 200. The beam has a hinge at B.

(a) Find the stiffness and carry-over factors of the beam.
(b) Calculate the fixed-end moments due to a concentrated load of 2 applied at C.

Answer: 5·45, 21·8
2·0, 0·5
−4·54, 10·92

3.4 A non-uniform beam connects two joints i and j in a structure. If it is rigidly attached to both joints as in fig. 3.16(a), it has stiffness factors

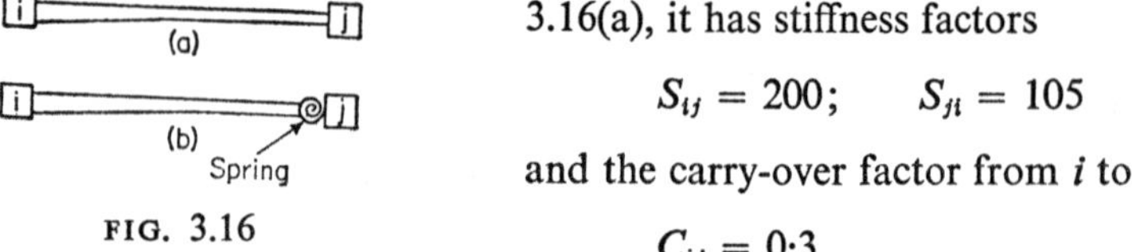

FIG. 3.16

$$S_{ij} = 200; \qquad S_{ji} = 105$$

and the carry-over factor from i to j is

$$C_{ij} = 0{\cdot}3$$

If a rotational spring of stiffness 100 units is now inserted between the end of the beam and joint j, as in fig. 3.16(b), such that only a relative rotation but no linear displacement can occur between the beam and the joint, find the new stiffness factor S'_{ij} and carry-over factor C'_{ij} for the beam-spring combination.

Answer: 185·6, 0·129

3.5 A non-uniform beam AB has a length of 20 and is haunched at A. Tables give

$$S_{AB} = \frac{6{\cdot}38EI}{L}; \qquad C_{AB} = 0{\cdot}382$$

$$S_{BA} = \frac{4{\cdot}56EI}{L}; \qquad C_{BA} = 0{\cdot}535$$

for such a beam with both ends fixed.

Find the fixed-end moment at A if the beam is simply supported (pinned) at B and a vertical concentrated load of 15 is applied at a point 8 units from B. If the beam were simply supported the end rotations would be $\theta_A = 0{\cdot}024PL^2/EI$ and $\theta_B = -\ 0{\cdot}055PL^2/EI$.

Answer: $-36{\cdot}6$.

4

Flexibility and Stiffness Matrices

4.1 The Advantages of a Matrix Formulation

For a structure with many node points and with many discrete forces and displacements involved in its analysis, it is often easier to use a matrix formulation to deal with whole groups of forces or displacements in a single operation rather than to treat them individually. Such a procedure has a number of advantages, some of which are:

(a) Matrix formulations are helpful in the understanding of basic theoretical concepts in that fundamental principles can more readily be seen when they are not hidden in a confusion of individual numbers, quantities and operations. The individual numbers are not, of course, lost, for a matrix operation can be halted at any stage so that specific figures may be examined: however, these figures are never allowed to obscure the overall process of analysis.

(b) The use of matrices is often a practical help in systematising computations.

(c) Because of the inherent logic of matrix operations, matrix methods form the basis of computer programs for structural analysis.

It must be admitted, however, that it is generally inefficient to use matrix procedures alone for computation, either by hand or with the use of a computer, for the very rigour of the systematisation which is so helpful in formulating procedures of analysis and solution tends to preclude the use of short cuts in calculation and of efficient storage use in a computer.

This chapter seeks only to introduce the concepts of flexibility and stiffness matrices: a systematic discussion of matrix methods of structural analysis is left until Chapter 8, although matrices do occur in various contexts throughout the rest of the book.

4.2 Flexibility Matrices

Fig. 4.1 represents a typical structure. If a force P_1 is applied to it, the whole structure will deflect and points 1 and 2, for example, will be displaced by amounts Δ_1 and Δ_2. There will be some relationship between the force and the displacements, which could be expressed as

$$\Delta_1 = f_{11}P_1$$
$$\Delta_2 = f_{21}P_1$$

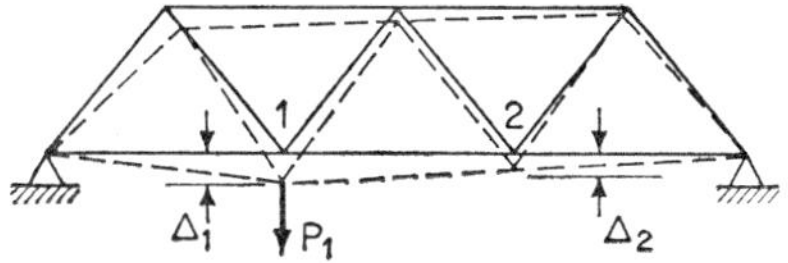

FIG. 4.1

where f_{11} and f_{21} are coefficients depending on the geometry and elastic (or inelastic) properties of the structure. They are called *flexibility coefficients.*

There is little point in using flexibility coefficients except for a linear elastic structure for which the principle of elastic superposition holds. In such a system the flexibility coefficient f_{ij} may be defined as the deflection at i due to a unit force at j, and in addition, the total displacement of a point due to a number of forces (fig. 4.2(a)) could be expressed as the sum of the displacement due to each force separately, or

$$\Delta_1 = f_{11}P_1 + f_{12}P_2 + f_{13}P_3 + f_{14}P_4 \tag{4.1(a)}$$

It should be noted that each number used as a subscript implies not only a point at which a force is applied or a displacement occurs, but also a direction at that point. Thus in fig. 4.2(b), 1 refers to an inclined direction at the left-hand joint while 2 and 3 refer to the vertical and horizontal directions at the top of the frame. It should also be noted that matrix analysis, with certain exceptions occurring in more advanced analysis, is concerned with the displacements and forces at only a finite number of points on a structure and not with continuous distributions.

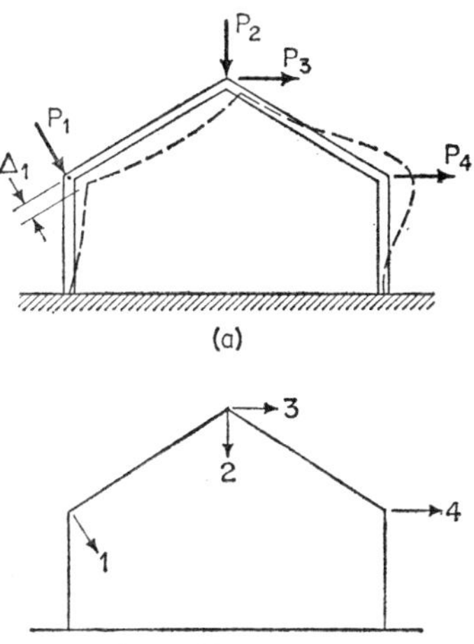

FIG. 4.2

If displacements are measured at each of the point-directions of application of the forces, then we could also write

$$\left.\begin{aligned}\Delta_2 &= f_{21}P_1 + f_{22}P_2 + f_{23}P_3 + f_{24}P_4\\ \Delta_3 &= f_{31}P_1 + f_{32}P_2 + f_{33}P_3 + f_{34}P_4\\ \Delta_4 &= f_{41}P_1 + f_{42}P_2 + f_{43}P_3 + f_{44}P_4\end{aligned}\right\} \tag{4.1(b)}$$

Equations (4.1) could readily be expressed in matrix form as

$$\begin{Bmatrix} \Delta_1 \\ \Delta_2 \\ \Delta_3 \\ \Delta_4 \end{Bmatrix} = \begin{bmatrix} f_{11} & f_{12} & f_{13} & f_{14} \\ f_{21} & f_{22} & f_{23} & f_{24} \\ f_{31} & f_{32} & f_{33} & f_{34} \\ f_{42} & f_{42} & f_{43} & f_{44} \end{bmatrix} \begin{Bmatrix} P_1 \\ P_2 \\ P_3 \\ P_4 \end{Bmatrix} \tag{4.2}$$

and using symbols for the matrices, this becomes

$$\mathbf{\Delta} = \mathbf{FP} \tag{4.3}$$

F is known as a *flexibility matrix*, while **Δ** and **P** are column vectors of displacements and forces. The number of forces is the same as the number of displacements so that a flexibility matrix is always square, and the displacements must always be written in the same order as the corresponding forces to preserve the symmetry of coefficients shown in equation (4.2). It should be

FIG. 4.3

FIG. 4.4

remembered that the words 'force' and 'displacement' are used in a generalised sense to represent not only linear forces and displacements but also moments and rotational displacements. In fact, flexibility matrices can be derived in terms of generalised forces and displacements, where each generalised force and displacement represents a whole group of dependent single forces and displacements. However, such concepts are beyond the scope of this book.

If the uniform cantilever of fig. 4.3 is subjected to a force P_1 in direction 1, it will extend by an amount P_1L/EA but will not rotate or deflect transversely. Hence three of the required nine flexibility coefficients for point-directions 1, 2 and 3 will be

$$f_{11} = \frac{L}{EA}$$

$$f_{21} = f_{31} = 0$$

and the first column of the flexibility matrix is obtained. From Table 1.1 on page 26 the transverse displacement and the rotation due to a transverse force P_2 will be $P_2L^3/3EI$ and $P_2L^2/2EI$ respectively, while the axial displacement is zero. Hence three more of the required coefficients are

$$f_{12} = 0; \qquad f_{22} = \frac{L^3}{3EI}; \qquad f_{32} = \frac{L^2}{2EI}$$

Again from Table 1.1 the transverse and rotational displacements due to a moment M_3 will be $M_3L^2/2EI$ and M_3L/EI while the axial displacement is zero, so that

$$f_{13} = 0; \qquad f_{23} = \frac{L^2}{2EI}; \qquad f_{33} = \frac{L}{EI}$$

The relationship between the three forces and the three displacements may now be written as

$$\begin{bmatrix} L/EA & 0 & 0 \\ 0 & L^3/3EI & L^2/2EI \\ 0 & L^2/2EI & L/EI \end{bmatrix} \begin{Bmatrix} P_1 \\ P_2 \\ M_3 \end{Bmatrix} = \begin{Bmatrix} \Delta_1 \\ \Delta_2 \\ \theta_3 \end{Bmatrix} \tag{4.4}$$

or again

$$\mathbf{FP} = \mathbf{\Delta}$$

The flexibility coefficients relating forces and displacements at points 1 and 2 of the non-uniform cantilever of fig. 4.4 may be obtained by superposition of the effects of the two parts of the structure. We obtain

$$f_{11} = \frac{L^3}{6EI}$$

$$f_{12} = \frac{L^3}{6EI} + \frac{L^3}{4EI} = \frac{5L^3}{12EI}$$

$$f_{22} = \frac{5L^3}{12EI} + \frac{L^3}{4EI} + \frac{L^3}{2EI} + \frac{L^3}{3EI} = \frac{3L^3}{2EI}$$

$$f_{21} = \frac{L^3}{6EI} + \frac{L^3}{4EI} = \frac{5L^3}{12EI}$$

The flexibility matrix for the two points is therefore

$$\mathbf{F} = \begin{bmatrix} L^3/6EI & 5L^3/12EI \\ 5L^3/12EI & 3L^3/2EI \end{bmatrix} \tag{4.5}$$

As a further example, consider the pin-jointed truss of fig. 4.5. If a unit force is applied at 1, the resulting deflections are

$$f_{11} = \frac{a}{EA}$$

$$f_{21} = -\frac{a}{EA}$$

For a unit force applied at 2, the bar forces in AB and BC are $\sqrt{2}$ and -1 respectively, taking a tensile force as positive. Point B will move to B′, so that from the geometry of the displacement diagram (fig. 4.5(b)) we obtain

$$f_{22} = \frac{a}{EA} + \sqrt{2} \times \sqrt{2} \times \frac{a\sqrt{2}}{EA} = 3{\cdot}828\frac{a}{EA}$$

$$f_{12} = -\frac{a}{EA}$$

Hence we can write

$$\mathbf{F} = \begin{bmatrix} a/EA & -a/EA \\ -a/EA & 3{\cdot}828a/EA \end{bmatrix} \tag{4.6}$$

Looking at the three equations (4.4), (4.5) and (4.6), two points may be noted about a flexibility matrix. First, while off-diagonal terms may be positive, negative or zero, terms on the leading diagonal of the matrix are always positive. They are bound to be positive, for if they were not, a force applied at a point would cause a displacement in the opposite direction to the force—an impossible situation. It will also be noticed that the matrices are all symmetrical. The reason for this is discussed in the next section.

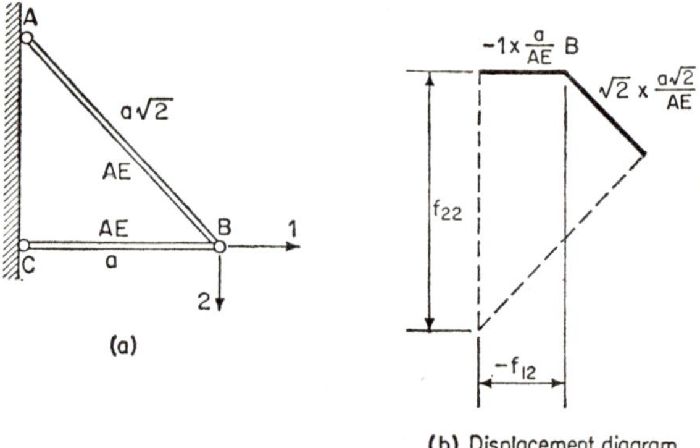

FIG. 4.5

4.3 Reciprocal Theorems

From the principle of elastic superposition we know that when a series of forces is applied to a linearly elastic structure the final deformation of the structure is independent of the order of application of the forces. The strain energy in the structure must also be independent of the order of application of the forces, and so must the total work done on the structure.

Two forces are applied to the linearly elastic structure of fig. 4.6. If P_1 is applied first, causing displacements Δ_1' and Δ_2', then the work done is

$$W' = \tfrac{1}{2}P_1\Delta_1' = \tfrac{1}{2}P_1 f_{11} P_1$$

If P_2 is now applied causing additional displacements Δ_1'' and Δ_2'', then the work done by P_1 moving through Δ_1'' is

$$W'' = P_1\Delta_1'' = P_1 f_{12} P_2$$

and the work done by P_2 is

$$W''' = \tfrac{1}{2}P_2\Delta_2'' = \tfrac{1}{2}P_2 f_{22} P_2$$

The total work done on the structure is thus

$$\begin{aligned} W &= W' + W'' + W''' \\ &= \tfrac{1}{2}P_1 f_{11} P_1 + P_1 f_{12} P_2 + \tfrac{1}{2}P_2 f_{22} P_2 \end{aligned}$$

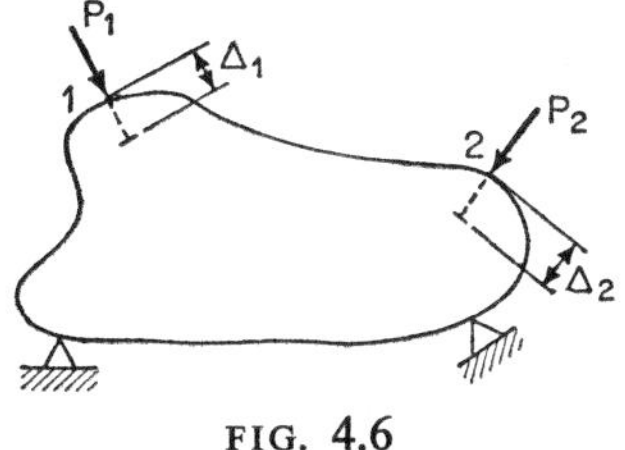

FIG. 4.6

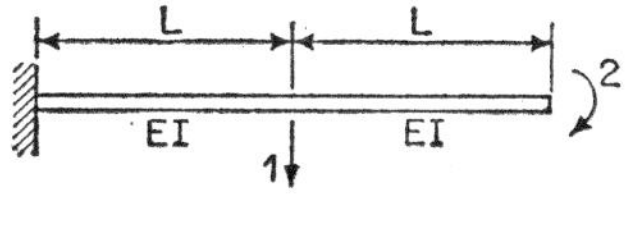

FIG. 4.7

If, however, the two forces were applied in the opposite order, then the work done would be

$$W = \tfrac{1}{2}P_2 f_{22} P_2 + P_2 f_{21} P_1 + \tfrac{1}{2}P_1 f_{11} P_1$$

As the work done is the same irrespective of the order of loading, then

$$f_{12} = f_{21}$$

and in general when any number of forces and displacements are applied to a structure, for any pair of point-directions *i* and *j*,

$$f_{ij} = f_{ji} \tag{4.7}$$

and **F** is symmetric. This is an expression of *Maxwell's reciprocal theorem* which could also be stated:

> The displacement at *i* due to a unit force at *j* is equal to the displacement at *j* due to a unit force at *i*, for a linearly elastic system.

For the uniform cantilever shown in fig. 4.7, the rotation at 2 due to a unit force at 1 is

$$f_{21} = \frac{L^2}{2EI}$$

while the displacement at 1 due to a unit moment at 2 is also

$$f_{12} = \frac{L^2}{2EI}$$

A more generalised form of the reciprocal theorem may be obtained from Maxwell's theorem as follows. Consider a linearly elastic structure with a number of point-directions defined on it for which the flexibility matrix **F** is known, and consider two separate systems of forces $\mathbf{P}_A$ and $\mathbf{P}_B$ applied to the structure at the point-directions, producing corresponding sets of displacements $\mathbf{\Delta}_A$ and $\mathbf{\Delta}_B$. Any of the elements of $\mathbf{P}_A$ and $\mathbf{P}_B$ may, of course, be zero. Multiplying one set of forces by the displacements due to the other set and expanding,

$$\begin{aligned}
\mathbf{P}_A^T\mathbf{\Delta}_B &= \lfloor P_{A1}, P_{A2}, \ldots \rfloor \begin{Bmatrix} \Delta_{B1} \\ \Delta_{B2} \\ \vdots \end{Bmatrix} \\
&= \mathbf{P}_A^T\mathbf{F}\mathbf{P}_B \\
&= (\mathbf{P}_A^T\mathbf{F}\mathbf{P}_B)^T \quad \text{as it is a scalar} \\
&= \mathbf{P}_B^T\mathbf{F}^T\mathbf{P}_A \\
&= \mathbf{P}_B^T\mathbf{F}\mathbf{P}_A \quad \text{as } \mathbf{F} \text{ is symmetric} \\
&= \mathbf{P}_B^T\mathbf{\Delta}_A
\end{aligned}$$

Hence we can say that if two systems of forces are applied to a linearly elastic structure, then the sum of the products of the first set of forces with the corresponding displacements due to the second set is equal to the sum of the product of the second set of forces with the displacements due to the first. This is a statement of *Betti's reciprocal theorem*, of which Maxwell's theorem is a particular case.

4.4 Stiffness Matrices

Equation (4.2) expresses the displacements of a structure in terms of the forces applied to it. We could equally well write a converse relationship expressing forces in terms of displacements thus:

$$\begin{Bmatrix} P_1 \\ P_2 \\ P_3 \\ P_4 \end{Bmatrix} = \begin{bmatrix} k_{11} & k_{12} & k_{13} & k_{14} \\ k_{21} & k_{22} & k_{23} & k_{24} \\ k_{31} & k_{32} & k_{33} & k_{34} \\ k_{41} & k_{42} & k_{43} & k_{44} \end{bmatrix} \begin{Bmatrix} \Delta_1 \\ \Delta_2 \\ \Delta_3 \\ \Delta_4 \end{Bmatrix} \tag{4.8}$$

which could be written

$$\mathbf{P} = \mathbf{K}\boldsymbol{\Delta} \tag{4.9}$$

where the coefficients k_{ij} are called *stiffness coefficients* and **K** is called a *stiffness matrix*. Like a flexibility matrix and for the same reasons, a stiffness matrix is square and symmetric, and the terms on the leading diagonal are positive.

A typical flexibility coefficient f_{ij} was defined to be the displacement at i due to a unit force applied at j. The definition of a stiffness coefficient is analogous to this: it is that a typical coefficient k_{ij} represents the force at i due to a unit displacement applied at j. This is illustrated in fig. 4.8, where it

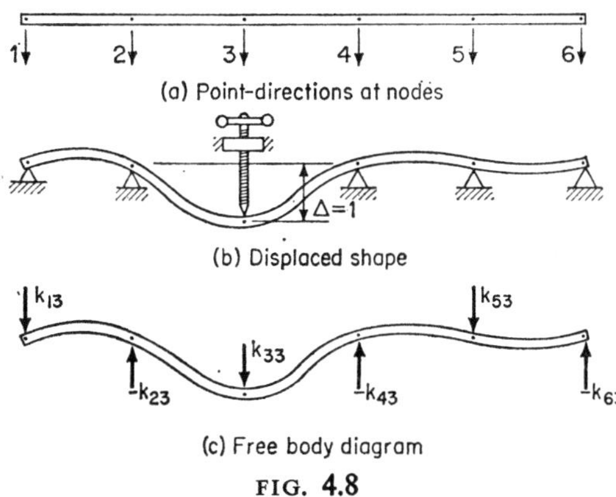

FIG. 4.8

should be noticed that only the node point (or point-direction) at which a unit displacement is applied is allowed to move, for all other displacements must be fixed at zero. This may be compared with the flexibility concept in which no force is applied to any of the nodes except that at which the unit force is specified.

It can be seen from equations (4.3) and (4.9) that the relationship between the stiffness and flexibility matrices of a structure is simply that the stiffness matrix is the inverse of the flexibility matrix, or

$$\mathbf{K} = \mathbf{F}^{-1} \tag{4.10}$$

The converse is usually true, that $\mathbf{F} = \mathbf{K}^{-1}$ except that stiffness matrices are sometimes formed which are singular and cannot be inverted. The reason for this is that to find the flexibility matrix for a given series of nodes on a structure, the structure must first be stable and sufficient reaction points must exist to bring the various unit forces applied to the nodes into equilibrium.

No equivalent requirement exists for a stiffness matrix. The beam of fig. 4.8 is an example in point: it is quite unattached to any foundation so that although there would be no difficulty in finding a stiffness matrix for the nodes shown, it would be impossible to put a single unit force on the structure, and no flexibility matrix could be obtained.

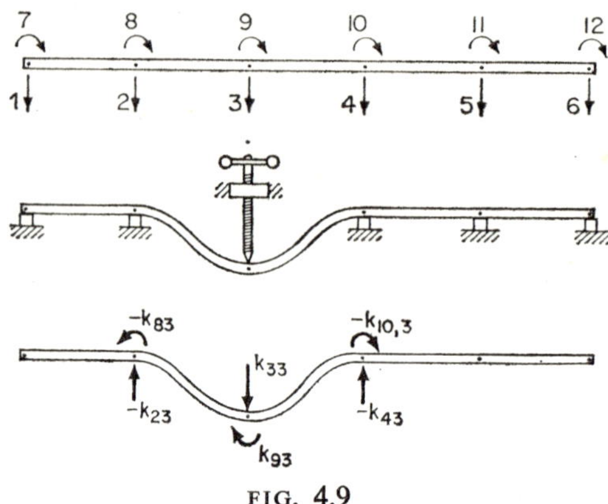

FIG. 4.9

In fig. 4.8, forces and displacements are only considered in the transverse direction, and it can be seen that the displacement of any one of the points will cause forces to occur at all of them. However, if moments and rotations are also taken into account as in fig. 4.9, then the displacement of, say, point 3 will produce forces at that point and at the adjacent nodes, but not at any other node. If neither linear nor rotational displacement can take place at the adjacent nodes, then no moment can be carried past these points and many of the stiffness coefficients will be zero. This illustrates a property of stiffness matrices not possessed by flexibility matrices, that if a stiffness matrix is formed for all the relevant degrees of freedom at the nodes of a structure, then the matrix will be sparse: that is, a great many of its terms will be zero. Because, too, a displacement at any node will only affect the adjacent nodes, it is much easier to form a stiffness matrix for a large structure than it is to form a flexibility matrix, provided, again, that all relevant degrees of freedom are considered at the nodes.

If a unit axial displacement is given to the uniform cantilever of fig. 4.10, then the axial force will be

$$k_{11} = \frac{EA}{L}$$

while, because there will be no moment or transverse force,

$$k_{21} = k_{31} = 0$$

If a transverse displacement is applied as in fig. 4.10(c), the deflected shape and the bending moment will be skew-symmetrical about the centre of the cantilever so that the bending moment there is zero. Hence treating half the beam as a cantilever with a shear of k_{22} applied to its end we may write from Table 1.1(a)

$$\frac{1}{2} = \frac{k_{22}(L/2)^3}{3EI}$$

so that

$$k_{22} = \frac{12EI}{L^3}$$

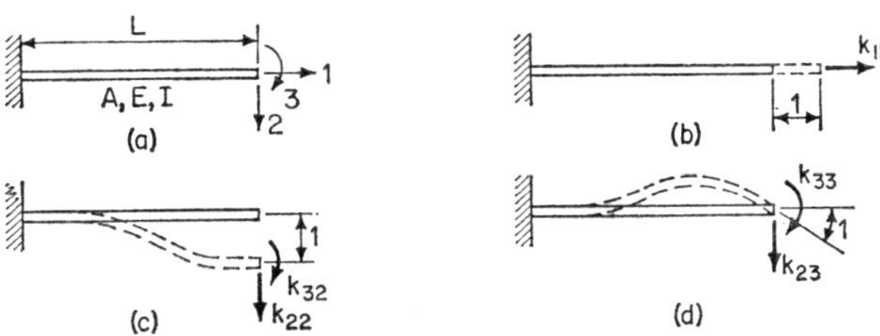

FIG. 4.10

Using statics, we find that

$$k_{32} = -\frac{L}{2}\cdot\frac{12EI}{L^3} = -\frac{6EI}{L^2}$$

From Table 3.1(a), for a unit rotation of the end of the cantilever the moment will be

$$k_{33} = \frac{4EI}{L}$$

and taking moments about the fixed end, remembering the moment there will be $\frac{1}{2}(4EI/L)$, we obtain

$$k_{23} = -\frac{6EI}{L^2}$$

The complete stiffness matrix for a uniform cantilever may now be written as

$$\mathbf{K} = \begin{bmatrix} EA/L & 0 & 0 \\ 0 & 12EI/L^3 & -6EI/L^2 \\ 0 & -6EI/L^2 & 4EI/L \end{bmatrix} \tag{4.11}$$

Fig. 4.11 shows a simply supported beam for which the rotations of the two ends are point-directions. It can be seen from the relevant definitions that the stiffness matrix may be written in terms of the stiffness and carry-over factors of the beam and will be

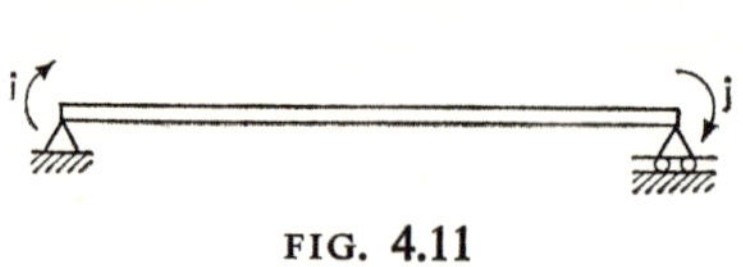

FIG. 4.11

$$\mathbf{K} = \begin{bmatrix} S_{ij} & C_{ji}S_{ji} \\ C_{ij}S_{ij} & S_{ji} \end{bmatrix} \tag{4.12}$$

Hence because of the symmetry of the matrix

$$C_{ij}S_{ij} = C_{ji}S_{ji} \tag{4.13}$$

This relationship was noted in the previous chapter because of its utility in checking calculated values of stiffness and carry-over factors.

4.5 A Further Example

To help fix ideas a more complicated example will now be calculated which is, to find the flexibility and stiffness matrices of the two-element structure shown in fig. 4.12, assuming deformations are small but including axial flexibility.

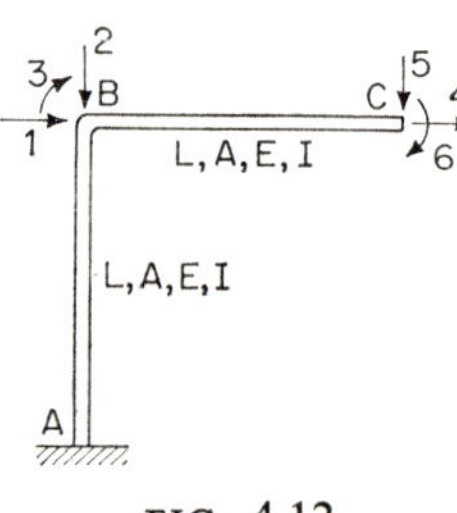

FIG. 4.12

The most straightforward way of calculating the flexibility matrix is to use the results given in Table 1.1 or by equation (4.4) together with the principles of superposition to find the effects of unit actions applied at each point-direction in turn. Beginning with a unit force applied at 1 (fig. 4.13(a)) we see that from Table 1.1(a) the displacement in direction 1 is

$$f_{11} = \frac{L^3}{3EI}$$

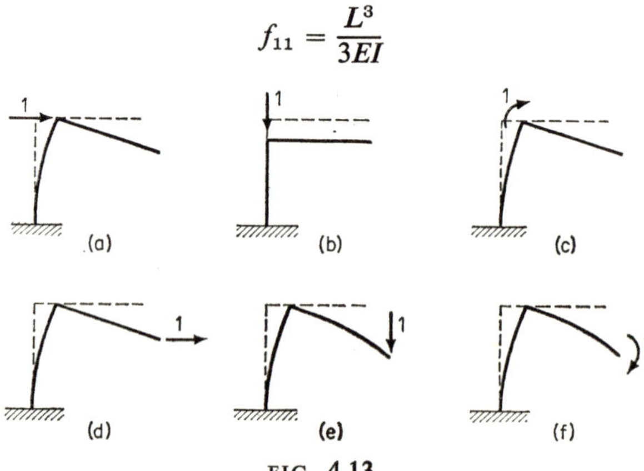

FIG. 4.13

Similarly,

$$f_{21} = 0 \quad \text{as the joint only moves horizontally}$$

$$f_{31} = \frac{L^2}{2EI}$$

$$f_{41} = \frac{L^3}{3EI} \quad \text{as it must be the same as } f_{11}$$

$$f_{51} = L \times f_{31} = \frac{L^3}{2EI}$$

$$f_{61} = f_{31} = \frac{L^2}{2EI}$$

The only effect of a unit force at 2 is to compress AB, so that

$$f_{22} = f_{52} = \frac{L}{EA}$$

$$f_{12} = f_{32} = f_{42} = f_{62} = 0$$

With a unit moment applied at 3,

$$f_{13} = \frac{L^2}{2EI}$$

$$f_{23} = 0 \quad \text{as the joint only moves horizontally}$$

$$f_{33} = \frac{L}{EI}$$

$$f_{43} = f_{13} = \frac{L^2}{2EI}$$

$$f_{53} = L \times f_{33} = \frac{L^2}{EI}$$

$$f_{63} = f_{33} = \frac{L}{EI}$$

A unit force applied at 4 has the same effect as one applied at 1, except that BC extends giving an extra displacement to 4. Hence

$$f_{14} = \frac{L^3}{3EI} \qquad f_{24} = 0$$

$$f_{34} = \frac{L^2}{2EI} \qquad f_{44} = \frac{L^3}{3EI} + \frac{L}{EA}$$

$$f_{54} = \frac{L^3}{2EI} \qquad f_{64} = \frac{L^2}{2EI}$$

A unit force at 5 will cause both AB and BC to deform. The total deformation of the structure may be considered to be made up of the three separate effects shown in fig. 4.14, in which

(a) AB is compressed by a unit force
(b) AB is bent by a moment L
(c) BC is bent by a unit transverse force.

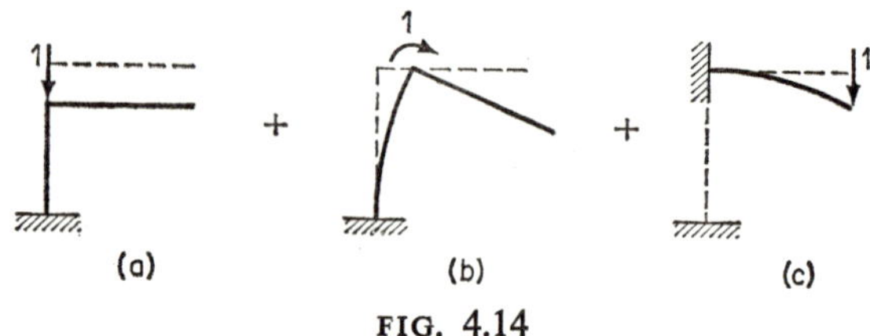

FIG. 4.14

Adding these effects, we get

$$
\begin{array}{lcccccccccc}
 & & \text{(a)} & & \text{(b)} & & \text{(c)} & & \text{TOTAL} \\
f_{15} & = & 0 & + & \dfrac{L^3}{2EI} & + & 0 & = & \dfrac{L^3}{2EI} \\
f_{25} & = & \dfrac{L}{EA} & + & 0 & + & 0 & = & \dfrac{L}{EA} \\
f_{35} & = & 0 & + & \dfrac{L^2}{EI} & + & 0 & = & \dfrac{L^2}{EI} \\
f_{45} & = & 0 & + & \dfrac{L^3}{2EI} & + & 0 & = & \dfrac{L^3}{2EI} \\
f_{55} & = & \dfrac{L}{EA} & + & \dfrac{L^3}{EI} & + & \dfrac{L^3}{3EI} & = & \dfrac{4L^3}{3EI} + \dfrac{L}{EA} \\
f_{65} & = & 0 & + & \dfrac{L^2}{EI} & + & \dfrac{L^2}{2EI} & = & \dfrac{3L^2}{2EI}
\end{array}
$$

The superposition approach used in obtaining these flexibility coefficients can be formalised by the use of certain multiplying matrices called transformation matrices, but a detailed discussion of this is left until Chapter 8.

With a unit moment at 6, both AB and BC again undergo deformations. Both members are subjected to a uniform bending moment of unity. The relevant flexibility coefficients may best be calculated by considering the deformation of AB and BC separately and adding the results, thus:

Displacement due to deformation of:

	AB		BC		TOTAL
$f_{16} =$	$\dfrac{L^2}{2EI}$	+	0	=	$\dfrac{L^2}{2EI}$
$f_{26} =$	0	+	0	=	0
$f_{36} =$	$\dfrac{L}{EI}$	+	0	=	$\dfrac{L}{EI}$
$f_{46} =$	$\dfrac{L^2}{2EI}$	+	0	=	$\dfrac{L^2}{2EI}$
$f_{56} =$	$\dfrac{L^2}{EI}$	+	$\dfrac{L^2}{2EI}$	=	$\dfrac{3L^2}{2EI}$
$f_{66} =$	$\dfrac{L}{EI}$	+	$\dfrac{L}{EI}$	=	$\dfrac{2L}{EI}$

Hence, the complete flexibility matrix is

$$\mathbf{F} = \begin{bmatrix} \dfrac{L^3}{3EI} & 0 & \dfrac{L^2}{2EI} & \dfrac{L^3}{3EI} & \dfrac{L^3}{2EI} & \dfrac{L^2}{2EI} \\ 0 & \dfrac{L}{EA} & 0 & 0 & \dfrac{L}{EA} & 0 \\ \dfrac{L^2}{2EI} & 0 & \dfrac{L}{EI} & \dfrac{L^2}{2EI} & \dfrac{L^2}{EI} & \dfrac{L}{EI} \\ \dfrac{L^3}{3EI} & 0 & \dfrac{L^2}{2EI} & \dfrac{L^3}{3EI}+\dfrac{L}{EA} & \dfrac{L^3}{2EI} & \dfrac{L^2}{2EI} \\ \dfrac{L^3}{2EI} & \dfrac{L}{EA} & \dfrac{L^2}{EI} & \dfrac{L^3}{2EI} & \dfrac{L}{EA}+\dfrac{4L^3}{3EI} & \dfrac{3L^2}{2EI} \\ \dfrac{L^2}{2EI} & 0 & \dfrac{L}{EI} & \dfrac{L^2}{2EI} & \dfrac{3L^2}{2EI} & \dfrac{2L}{EI} \end{bmatrix}$$

As a check we note that the matrix is square and symmetric, and that the terms on its leading diagonal are all positive.

The stiffness matrix of the structure may be found in a similar way by applying unit displacements at each point-direction in turn and calculating the actions they cause. With a unit displacement at 1 (and, of course, zero displacements at all other point-directions) the structure deforms as shown in fig. 4.15(a). AB bends and BC is compressed, so using the results given in equation (4.11) we obtain

$$k_{11} = \frac{12EI}{L^3} + \frac{EA}{L}$$

$$k_{31} = -\frac{6EI}{L^2}$$

$$k_{41} = -\frac{EA}{L}$$

$$k_{21} = k_{51} = k_{61} = 0$$

Note that careful attention must be paid to signs.

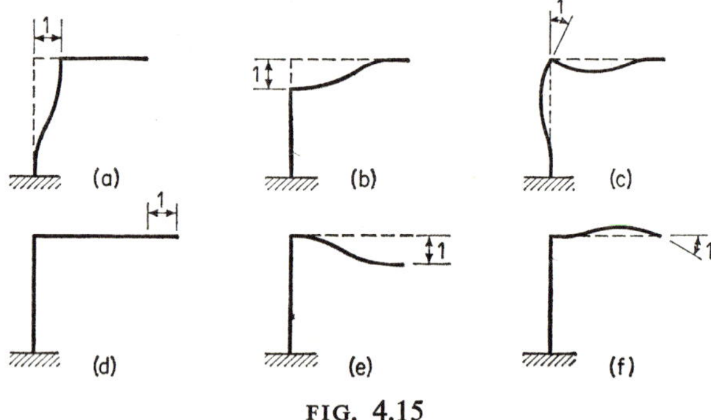

FIG. 4.15

The effect of a unit displacement at 2 (fig. 4.15(b)) is similar, except that this time BC bends and AB is compressed axially. The stiffness coefficients are therefore

$$k_{22} = \frac{12EI}{L^3} + \frac{EA}{L}$$

$$k_{32} = \frac{6EI}{L^2}$$

$$k_{52} = -\frac{12EI}{L^3}$$

$$k_{62} = \frac{6EI}{L^2}$$

$$k_{12} = k_{42} = 0$$

A unit rotation at 3 causes the deformed shape of fig. 4.15(c) in which it can be seen that the two members are bent into the same shape. The forces holding the structure in equilibrium must therefore be, again using equation (4.11) and also equation (4.12) for k_{63},

$$k_{13} = -\frac{6EI}{L^2} \qquad k_{23} = \frac{6EI}{L^2}$$

$$k_{33} = \frac{4EI}{L} + \frac{4EI}{L} = \frac{8EI}{L} \qquad k_{43} = 0$$

$$k_{53} = -\frac{6EI}{L^2} \qquad k_{63} = \frac{2EI}{L}$$

The only effect of a unit displacement at 4 is to stretch BC (fig. 4.15(d)) so that

$$k_{14} = -\frac{EA}{L} \qquad k_{44} = \frac{EA}{L}$$

$$k_{24} = k_{34} = k_{54} = k_{64} = 0$$

With a unit displacement at 5 and a unit rotation at 6, deformation is entirely confined to bending effects in BC, for no force or moment can be transferred past B into AB. Hence the relevant stiffness coefficients are

$$k_{15} = 0 \qquad k_{25} = -\frac{12EI}{L^3}$$

$$k_{35} = -\frac{6EI}{L^2} \qquad k_{45} = 0$$

$$k_{55} = \frac{12EI}{L^3} \qquad k_{65} = -\frac{6EI}{L^2}$$

$$k_{16} = 0 \qquad k_{26} = \frac{6EI}{L^2}$$

$$k_{36} = \frac{2EI}{L} \qquad k_{46} = 0$$

$$k_{56} = -\frac{6EI}{L} \qquad k_{66} = \frac{4EI}{L}$$

The complete stiffness matrix of the structure may now be written down as

$$\mathbf{K} = \begin{bmatrix} \frac{EA}{L} + \frac{12EI}{L^3} & 0 & -\frac{6EI}{L^2} & -\frac{EA}{L} & 0 & 0 \\ 0 & \frac{EA}{L} + \frac{12EI}{L^3} & \frac{6EI}{L^2} & 0 & -\frac{12EI}{L^3} & \frac{6EI}{L^2} \\ -\frac{6EI}{L^2} & \frac{6EI}{L^2} & \frac{8EI}{L} & 0 & -\frac{6EI}{L^2} & \frac{2EI}{L} \\ -\frac{EA}{L} & 0 & 0 & \frac{EA}{L} & 0 & 0 \\ 0 & -\frac{12EI}{L^3} & -\frac{6EI}{L^2} & 0 & \frac{12EI}{L^3} & -\frac{6EI}{L^2} \\ 0 & \frac{6EI}{L^2} & \frac{2EI}{L} & 0 & -\frac{6EI}{L^2} & \frac{4EI}{L} \end{bmatrix}$$

The matrix is symmetric as expected. If both the flexibility and the stiffness matrix are correct, it should also be possible to multiply them together to get a unit matrix; that is, the relation

$$\mathbf{FK} = \mathbf{I}$$

should hold. This check is left to the reader.

4.6 Matrices and the Solution of Indeterminate Structures

A more detailed account of the use of matrix methods in the solution of statically indeterminate structures will be given in Chapter 8. Briefly, however, the force method as already described in Chapter 2 first chooses certain forces in a redundant structure as unknowns, and uses conditions of geometric compatibility where the redundant forces occur to set up equations from which they may be found. Denoting all the redundants in a structure by the column vector $\mathbf{R}$ and the geometric incompatibilities by ∂, then if a flexibility matrix $\mathbf{F}$ can be obtained for the structure which relates $\mathbf{R}$ to ∂ by the equation

$$\mathbf{FR} = \partial$$

then the redundants can be found simply by solving the equation.

The displacement method, on the other hand, takes the nodal displacements of the structure as unknowns and solves a series of equilibrium equations to find them. The effects of the applied forces on the structure are introduced as statically equivalent forces at the nodes. Fixed-end moments are such statically equivalent forces. If they are denoted by the column vector **P** and the unknown nodal displacements by **Δ**, then if a stiffness matrix **K** can be found such that

$$\mathbf{P} = \mathbf{K}\boldsymbol{\Delta}$$

then the displacements may be found by solving the equation. This is a basic formulation of the displacement method. The solution can be expressed symbolically as

$$\boldsymbol{\Delta} = \mathbf{K}^{-1}\mathbf{P}$$

although the stiffness matrix in such a relation would seldom have to be inverted in practice as it is more efficient to solve the equations directly.

PROBLEMS

4.1 For a simply supported single span uniform beam of length L and section stiffness EI, find the flexibility and stiffness matrices for the two end rotations and moments only.

4.2 Use the information given in equations (4.4) and (4.11) to find the flexibility and stiffness matrices for the axial, transverse and rotational directions of the mid-point of a uniform beam of length L, built-in at both ends.

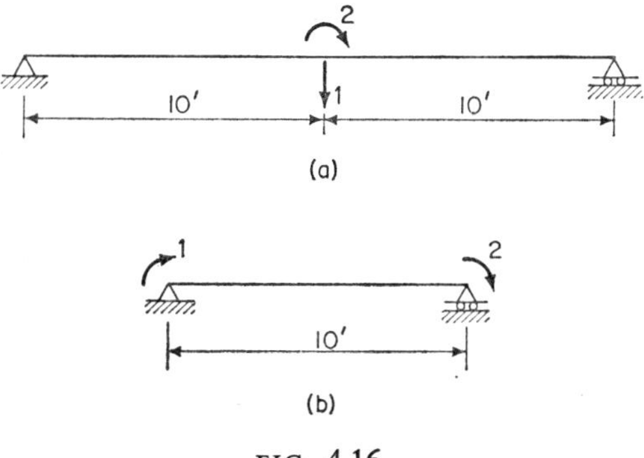

FIG. 4.16

4.3 Find the 2 × 2 flexibility matrix for the beam shown in fig. 4.16(a), given that the flexibility matrix for each of its components (fig. 4.16(b)) is

$$\mathbf{f} = \frac{1}{EI}\begin{bmatrix} 3{\cdot}3 & -1{\cdot}7 \\ -1{\cdot}7 & 3{\cdot}3 \end{bmatrix}$$

4.4 For the frame in fig. 4.17, find the flexibility matrix relating the forces and displacements in the

vertical direction at point 1
horizontal direction at point 2
vertical direction at point 3

Assume the members of the frame are infinitely stiff in their axial directions.

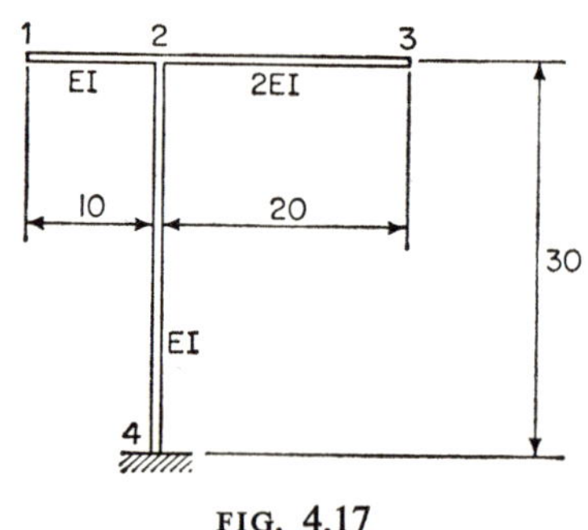

FIG. 4.17

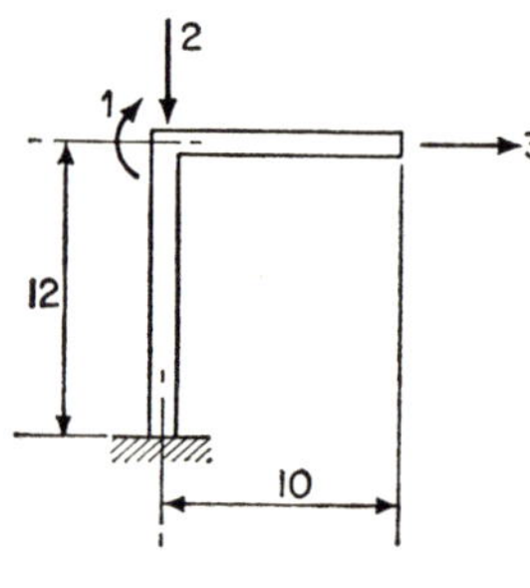

FIG. 4.18

4.5 Find the stiffness matrix and flexibility matrix for the three point-directions shown in fig. 4.18, and check your answer by forming their product.

For both members take the section bending and axial stiffnesses as

$$EI = 60 \times 10^7$$
$$EA = 3 \times 10^7$$

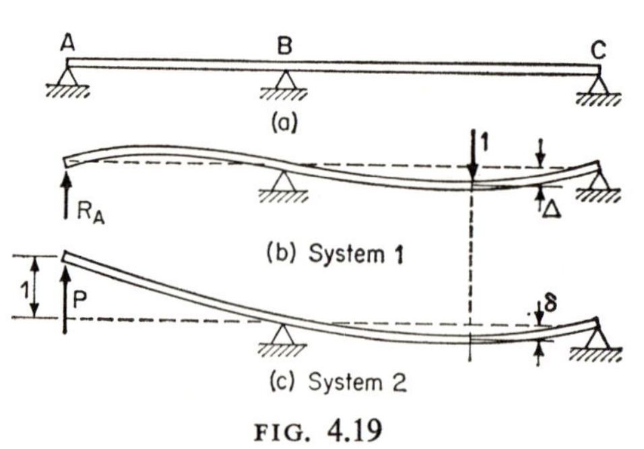

FIG. 4.19

4.6 The Müller – Breslau principle states in part that the influence line for a reaction of an elastic structure may be obtained by giving the structure a unit displacement at the point at which and in the direction in which the reaction occurs: the resulting deflected shape of the elastic line of the structure will then be the required influence line. Use Betti's reciprocal theorem to show that this is true for the two-span beam of fig. 4.19 by showing that the reaction R_A due to a unit load is equal to the displacement δ of the beam if A were given a unit vertical displacement. Use the two force and displacement systems given in fig. 4.19(b and c).

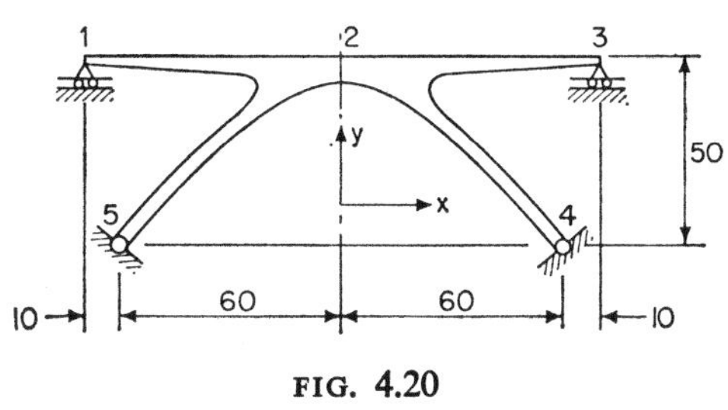

FIG. 4.20

In its complete form the Müller–Breslau principle also states that influence lines for internal forces in an elastic structure may be obtained by applying the equivalent internal displacements.

4.7 Fig. 4.20 shows a bridge frame. A model analysis gives the following results for the displacements at point 2 due to unit displacements at the support points 3, 4.

Unit displacement	Vertical displacement at 2	Horizontal displacement at 2
Vertical at 4	0·513	−0·385
Horizontal at 4	−0·183	0·500
Vertical at 3	−0·013	−0·027

All displacements are positive in the directions x, y. Use the Müller–Breslau principle

(a) to show that these results satisfy all the conditions of statics;

(b) to find the bending moment at point 2 due to a vertical load of 10 applied downwards at 2.

Answer: 207·4.

5

Force Methods of Frame Analysis

5.1 Introduction

The force method of analysis has already been described briefly in Section 2.7: the present chapter considers the method in more detail as it is applied to frame structures. Reference should be made at this stage to the summary given on page 46 which shows that the force method uses the kinematic restraints of a structure to solve for a set of unknown (redundant) forces.

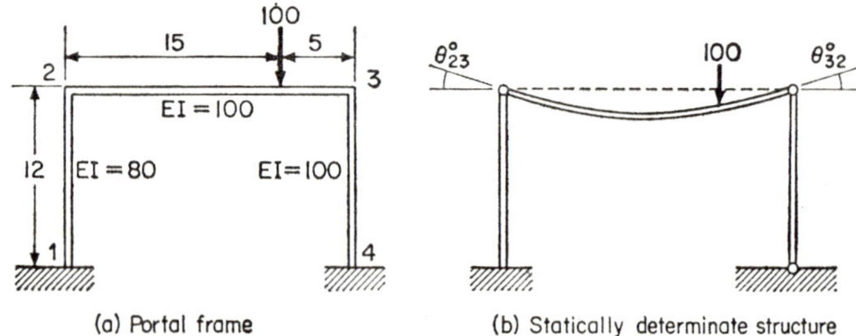

FIG. 5.1

A force method solution is in general much less easy to systematise than a displacement solution, because the range of choice of unknowns is greater and because there is usually no obvious choice of redundants. A force method analysis of a structure must therefore be preceded not only by an assessment of the degree of indeterminacy of the structure, but also by a decision as to which redundant forces to choose as unknowns. The two main factors influencing the decision are:

(a) ease of solution;
(b) accuracy of solution.

This point (among others) will now be illustrated by the solution of a portal frame using two different sets of unknowns.

5.2 Solution of a Portal Frame, I

The portal frame of fig. 5.1 is an asymmetric structure despite the symmetry of its geometry, because of the difference in the section stiffnesses *EI* of its two columns. The feet of the portal are built-in at 1 and 4 so that the structure is three times statically indeterminate and three unknown forces must be found. Choosing these as the bending moments at points 2, 3 and 4, the statically determinate structure will be as shown in fig. 5.1(b). The first step

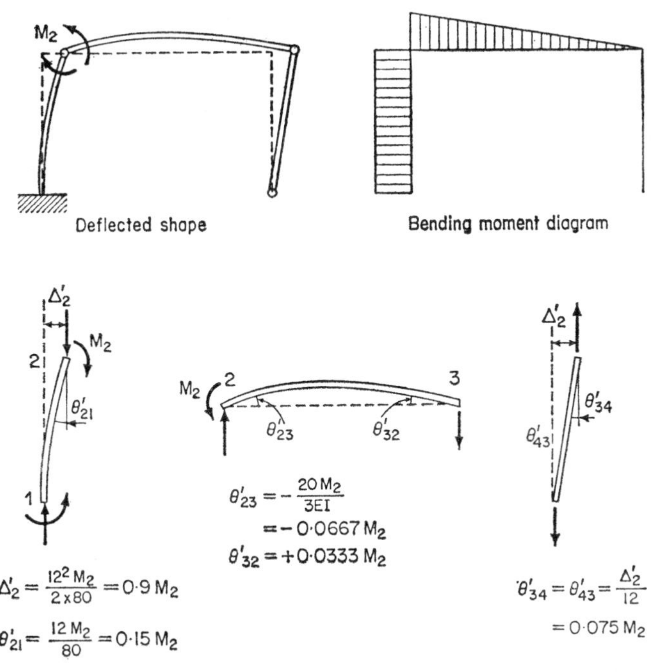

FIG. 5.2

in the solution is to calculate the geometric incompatibilities occurring in the statically determinate structure due to the applied load of 100. These are, that the ends of member 2–3 rotate through angles θ_{23}^0 and θ_{32}^0. Adopting the sign convention that clockwise rotations are positive, then by Table 1.1(d) (or the use of, say, the moment-area method)

$$\theta_{23}^0 = \frac{100 \times 15 \times 5 \times 25}{6 \times 20 \times 100} = +15{\cdot}63$$

$$\theta_{32}^0 = -\frac{100 \times 15 \times 5 \times 35}{6 \times 20 \times 100} = -21{\cdot}9$$

The three unknown moments, M_2, M_3 and M_4, are now applied to the statically determinate structure one at a time, with equal and opposite moments applied at either side of the relevant hinge to open or close the angle at the joint. The rotations of the ends of the members caused by the unknown moments are calculated in figs. 5.2, 5.3 and 5.4.

There are three unknowns, so that three compatibility equations are required. These are based on the condition that after the application of the applied load and the unknown moments, the members meeting at joints 2 and 3 must be at right angles to one another and member 3–4 must be

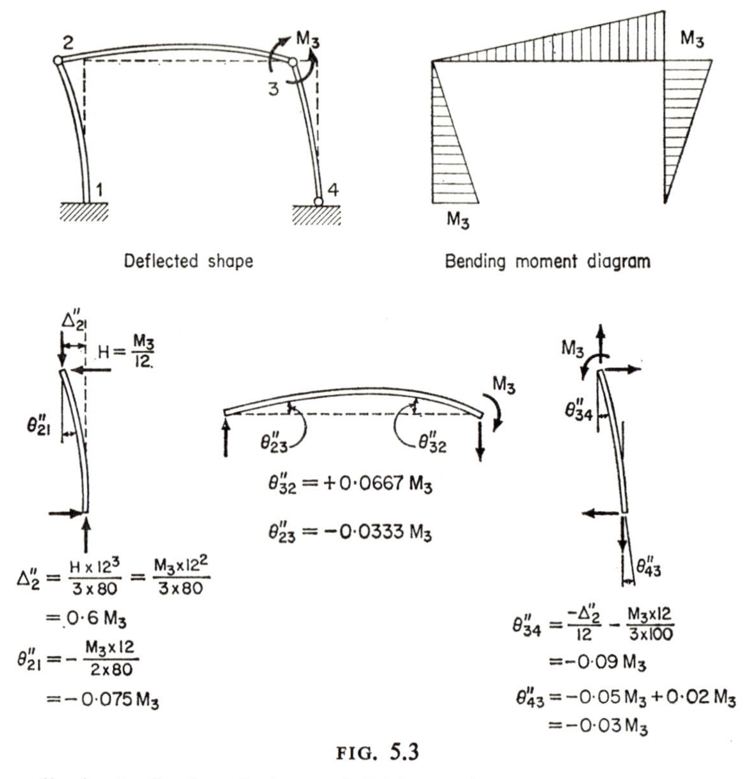

FIG. 5.3

perpendicular to the foundation at 4. If the total rotation of, for example, end 2 of member 2–3 is θ_{23}, then remembering that all rotations are taken to be positive clockwise, the compatibility equations will be

$$\begin{aligned} \theta_{21} - \theta_{23} &= 0 \\ \theta_{32} - \theta_{34} &= 0 \\ \theta_{43} \qquad &= 0 \end{aligned}$$

These may be expanded into the form

$$\begin{aligned} \theta'_{21} + \theta''_{21} + \theta'''_{21} - \theta'_{23} - \theta''_{23} - \theta'''_{23} &= \theta^0_{23} \\ \theta'_{32} + \theta''_{32} + \theta'''_{32} - \theta'_{34} - \theta''_{34} - \theta'''_{34} &= -\theta^0_{32} \\ \theta'_{43} + \theta''_{43} + \theta'''_{43} \qquad\qquad\qquad &= 0 \end{aligned}$$

where the primed rotations are caused by the unknown moments and are defined in figs. 5.2–5.4. Using the moment–rotation relationships found in these figures, the compatibility equations may now be written in terms of the unknown moments as

$$\begin{aligned} 0{\cdot}15M_2 - 0{\cdot}075M_3 + 0{\cdot}075M_4 + 0{\cdot}0667M_2 + 0{\cdot}0333M_3 &= 15{\cdot}63 \\ 0{\cdot}0333M_2 + 0{\cdot}0667M_3 - 0{\cdot}075M_2 + 0{\cdot}09M_3 - 0{\cdot}03M_4 &= 21{\cdot}9 \\ 0{\cdot}075M_2 - 0{\cdot}03M_3 + 0{\cdot}09M_4 &= 0 \end{aligned}$$

or

$$\begin{aligned} 0{\cdot}2167M_2 - 0{\cdot}0417M_3 + 0{\cdot}075M_4 &= 15{\cdot}63 \\ -0{\cdot}0417M_2 + 0{\cdot}1567M_3 - 0{\cdot}03M_4 &= 21{\cdot}9 \\ 0{\cdot}075M_2 - 0{\cdot}03M_3 + 0{\cdot}09M_4 &= 0 \end{aligned}$$

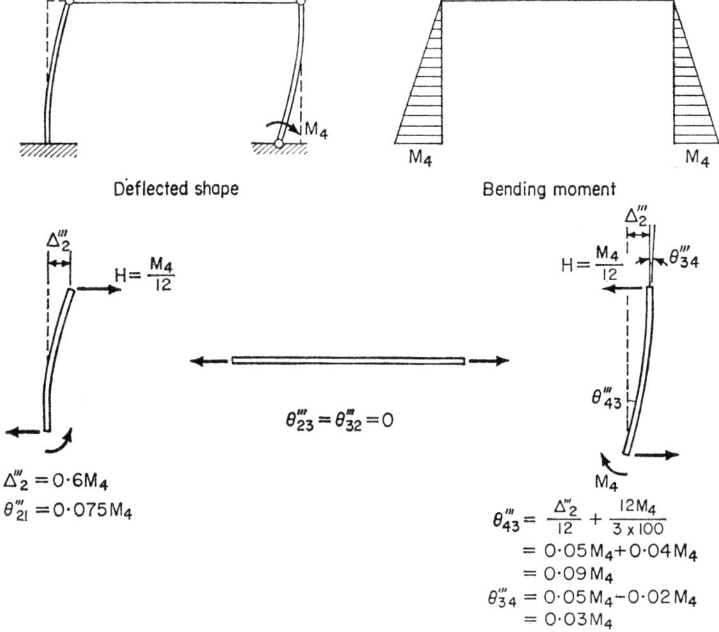

FIG. 5.4

Writing this in matrix form we have

$$\begin{bmatrix} 0{\cdot}2167 & -0{\cdot}0417 & 0{\cdot}075 \\ -0{\cdot}0417 & 0{\cdot}1567 & -0{\cdot}03 \\ 0{\cdot}075 & -0{\cdot}03 & 0{\cdot}09 \end{bmatrix} \begin{Bmatrix} M_2 \\ M_3 \\ M_4 \end{Bmatrix} = \begin{Bmatrix} 15{\cdot}63 \\ 21{\cdot}9 \\ 0 \end{Bmatrix} \tag{5.1}$$

The matrix of the coefficients of the moments is square and symmetrical: it is in fact the flexibility matrix relating the joint moments to the relative rotations of the members at the joints.

The equations may be solved by Gaussian elimination as follows. First, the flexibility matrix of equation (5.1) is augmented by the right-hand side leading to

$$\begin{bmatrix} 0{\cdot}2167 & -0{\cdot}0417 & 0{\cdot}075 & 15{\cdot}63 \\ -0{\cdot}0417 & 0{\cdot}1567 & -0{\cdot}03 & 21{\cdot}9 \\ 0{\cdot}075 & -0{\cdot}03 & 0{\cdot}09 & 0 \end{bmatrix}$$

Each row is now divided by its first term to give

$$\begin{bmatrix} 1 & -0{\cdot}1922 & 0{\cdot}346 & 72{\cdot}1 \\ 1 & -3{\cdot}755 & 0{\cdot}720 & -525 \\ 1 & -0{\cdot}400 & 1{\cdot}200 & 0 \end{bmatrix}$$

Subtracting the first line from the second two, we obtain

$$\begin{bmatrix} -3{\cdot}563 & 0{\cdot}374 & -597 \\ -0{\cdot}208 & 0{\cdot}854 & -72{\cdot}1 \end{bmatrix}$$

Dividing again by the first term in each row,

$$\begin{bmatrix} 1 & -0{\cdot}1050 & +167{\cdot}5 \\ 1 & -4{\cdot}10 & +346{\cdot}5 \end{bmatrix}$$

Subtracting,

$$[-4{\cdot}00 \quad +179{\cdot}0]$$

which represents the equation

$$-4{\cdot}00M_4 = +179$$

Hence

$$M_4 = 44{\cdot}8$$

Back-substituting to find the other moments

$$M_3 + 0{\cdot}105 \times 44{\cdot}8 = 167{\cdot}5$$
$$\therefore\ M_3 = 167{\cdot}5 - 4{\cdot}7 = 162{\cdot}8$$
$$M_2 - 0{\cdot}1922 \times 162{\cdot}8 - 0{\cdot}346 \times 44{\cdot}8 = 72{\cdot}1$$
$$\therefore\ M_2 = 72{\cdot}1 + 31{\cdot}2 + 15{\cdot}5 = 118{\cdot}8$$

Once the three unknown joint bending moments are known, the structure is statically determinate and all other reactions and internal forces can be found by statics: it can easily be shown, for instance, that the bending moment at joint 1 is 88·8, and the bending moment at the point of application of the load is 223. The bending moment diagram for the structure may now be drawn, as in fig. 5.5. This is an important step as it provides a basic check of the solution. An engineer will soon acquire sufficient experience to know whether a bending moment diagram looks approximately correct; but in any

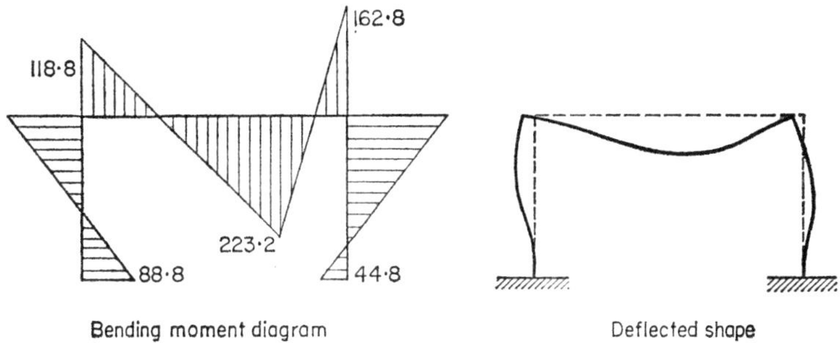

FIG. 5.5

case the bending moment diagram can be used (with the application of a little judgement) to construct a diagram of the approximate deflected shape of the structure using the fact that the tension side of a deforming member must be on the same side of the member as the bending moment diagram at that point. As any structure must fulfil both conditions of equilibrium and conditions of compatibility, then if it is impossible to draw a displaced shape which is compatible with the bending moment diagram, the latter must be in error. Such a procedure cannot, of course, check the detailed accuracy of a solution, but it is a useful guard against gross errors.

5.3 Solution of a Portal Frame, II

The portal frame of fig. 5.1 will now be analysed using the quite different set of unknown forces shown in fig. 5.6, namely the three reactions V, H and M_4

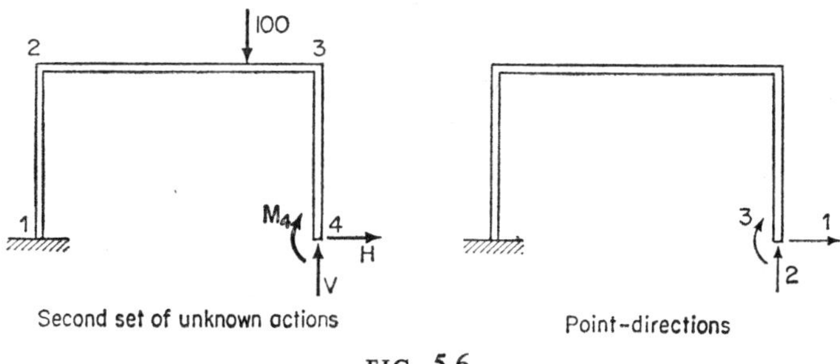

FIG. 5.6

at the right-hand foot of the frame. The frame is made statically determinate by completely cutting it at joint 4 so that the whole structure is cantilevered from the built-in foot at 1.

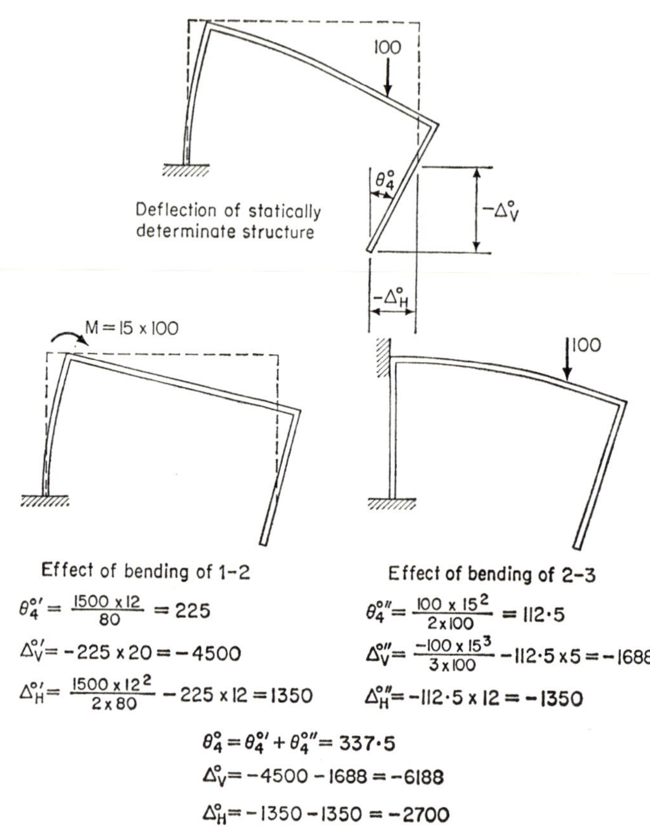

FIG. 5.7 Deflections due to applied load

The first step in the analysis is to find the deflections at point 4 of the statically determinate frame due to the application of the applied load. The calculations are carried out in fig. 5.7, where it will be seen that a superposition approach is used and that the displacements at 4 due to the deformation of members 1–2 and 2–3 are calculated separately then added. This type of approach has the advantage that the calculations as a whole are divided into a large number of separate and often trivial individual calculations: the method is dealt with more formally in matrix terms in Chapter 8.

In the same way the horizontal, vertical and rotational displacements of 4 are calculated in figs. 5.8–5.10 for each of the three unknown actions in turn by superimposing the separate displacement effects at 4 due to members 1–2, 2–3 and 3–4. It can be seen that unit forces rather than the unknown forces

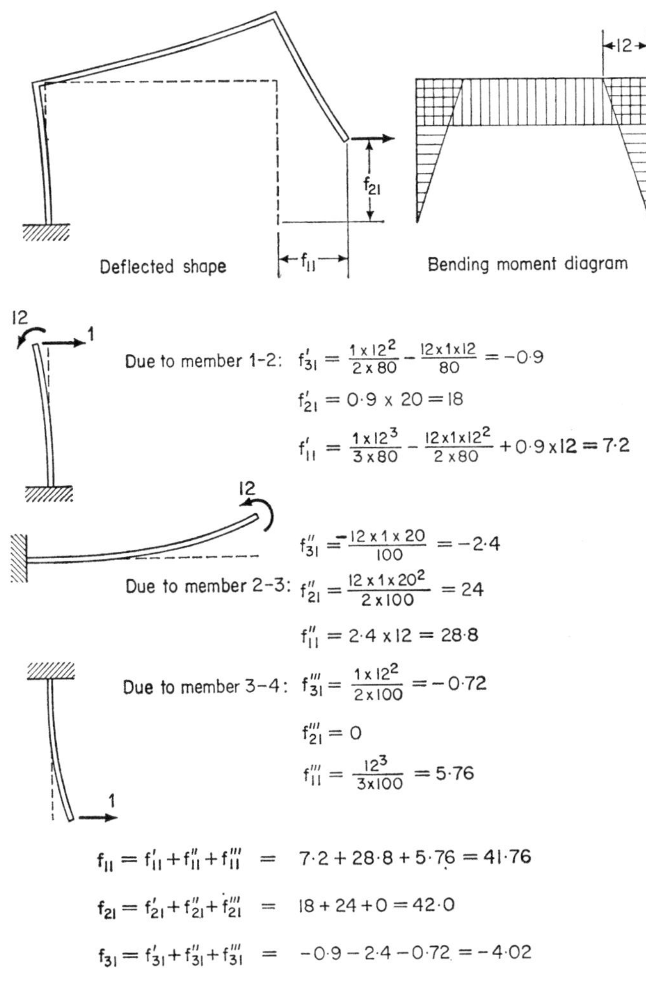

FIG. 5.8 Deflections due to horizontal force

have been applied at 4: the deflections at 4 are then simply the flexibility coefficients for the horizontal, vertical and angular directions at that point so that the flexibility coefficient notation adopted in the previous chapter has been used.

The compatibility conditions are, that point 4 is fixed and can neither rotate nor move horizontally or vertically. The displacements due to the

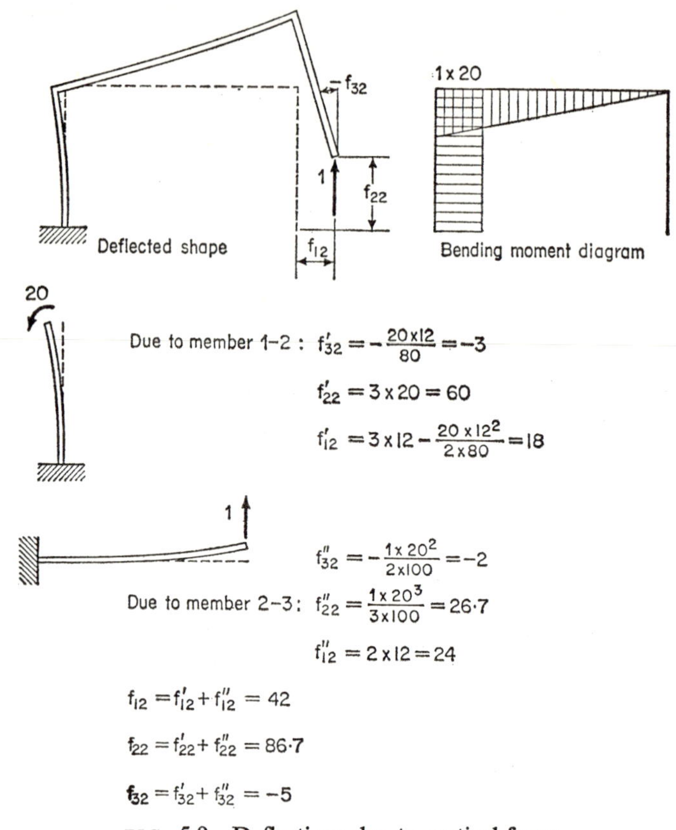

FIG. 5.9 Deflections due to vertical force

restoring forces must be equal to the negative of the displacements of the statically determinate structure at 4 due to the applied load, which leads to the equations

$$\begin{aligned} f_{11}H + f_{12}V + f_{13}M_4 &= 2700 \\ f_{21}H + f_{22}V + f_{23}M_4 &= 6188 \\ f_{31}H + f_{32}V + f_{33}M_4 &= -337{\cdot}5 \end{aligned}$$

which can, of course, be written in matrix terms as

$$\begin{bmatrix} 41{\cdot}76 & 42{\cdot}0 & -4{\cdot}02 \\ 42{\cdot}0 & 86{\cdot}7 & -5{\cdot}0 \\ -4{\cdot}02 & -5{\cdot}0 & 0{\cdot}47 \end{bmatrix} \begin{Bmatrix} H \\ V \\ M_4 \end{Bmatrix} = \begin{Bmatrix} 2700 \\ 6188 \\ -337{\cdot}5 \end{Bmatrix} \tag{5.2}$$

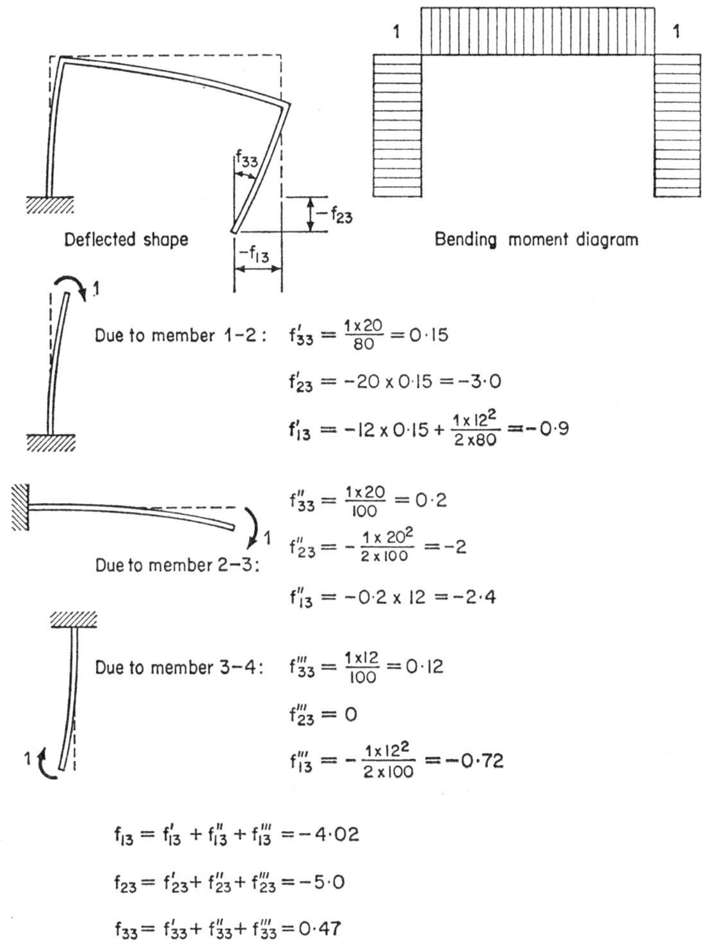

FIG. 5.10 Deflection due to moment

where the appropriate values of the flexibility coefficients have been used. The solution of this equation gives, to slide-rule accuracy,

$$H = -17{\cdot}5$$
$$V = 77{\cdot}2$$
$$M_4 = -46$$

and from these results the bending moments at any point in the frame may be obtained by statics.

5.4 Accuracy of Solution

The accuracy to which a set of simultaneous equations may be solved can be determined roughly by considering its coefficient matrix. The more the diagonal term predominates in each row or column, the more accurate will be the solution, whereas if the diagonal term is smaller than some of the other terms in the same row or column, or is even of the same order, then the solution will involve small differences between large numbers and its accuracy will be reduced. When the diagonal terms are strongly predominant the set of simultaneous equations is said to be *well conditioned*, while if it is difficult to obtain an accurate solution the equations are called *ill conditioned.* Comparing equations (5.1) and (5.2), it can be seen that the first set of equations is fairly well conditioned as the first two diagonal terms in the flexibility matrix are predominant and the third is still the largest element in its row and column. However, in equation (5.2), only f_{22} is predominant in its row and column: f_{11} is of the same order as f_{12}, while in the last row and column the diagonal element is by far the smallest term. Equation (5.1) would therefore be expected to give a more accurate solution than equation (5.2).

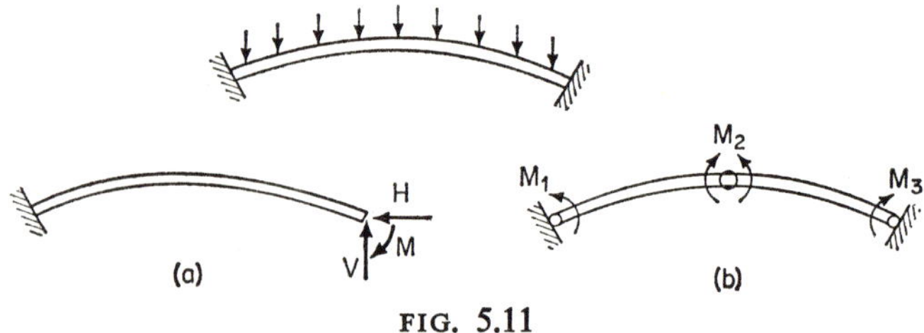

FIG. 5.11

It can be seen, therefore, that for any given structure the conditioning of the flexibility matrix depends on the choice of redundants in the structure, some choices leading to better conditioned equations than others. The best choice of unknowns to minimise the off-diagonal terms of the flexibility matrix will obviously be when each unknown force produces only small deflections at point-directions other than the one at which the force is applied. This can generally be achieved by choosing secondary effects as the redundants wherever possible such that behaviour of the statically determinate structure approximates fairly closely to that of the complete structure. Fig. 5.11 shows the two choices (a) and (b) of the redundant forces for an arch. The predominant action of the arch will be to resist the applied load by direct thrust so that the bending moments in such a structure will be secondary effects. A choice of the three bending moments in (b) will lead to a better conditioned set of equations than a choice of the set of forces in (a), for if the set (a) is chosen, the horizontal force, for instance, will cause large vertical and rotational displacements.

A choice of secondary effects as unknowns also helps the accuracy of the final solution in that a given percentage error in a redundant of this type will have far less effect on the internal force distribution in a structure than the same error in a primary effect.

Looking at the problem in a slightly different way, the redundant forces in a structure may be thought of as removing the compatibility errors existing in the statically determinate structure. If these errors are large, as in fig. 5.11(a), any inaccuracy in the calculation of the redundants will seriously effect the behaviour of the structure, so that the choice of unknowns should be such that the compatibility errors are as small as possible.

This principle could be applied to the solution of the portal frame given in the previous section by choosing the statically determinate structure shown in fig. 5.12. Initial estimates have been made of the vertical and horizontal forces at point 4: the vertical force of 75 units is obtained by taking member 2–3 as a simply supported beam and calculating the reaction at 3, while the horizontal force is merely estimated, from experience, at 10% of the applied load. The moment at 4 is taken to be zero at this stage. If now H' and V' are the extra forces required to be added to the initial estimates to give the true reactions, then equations (5.2) can be written

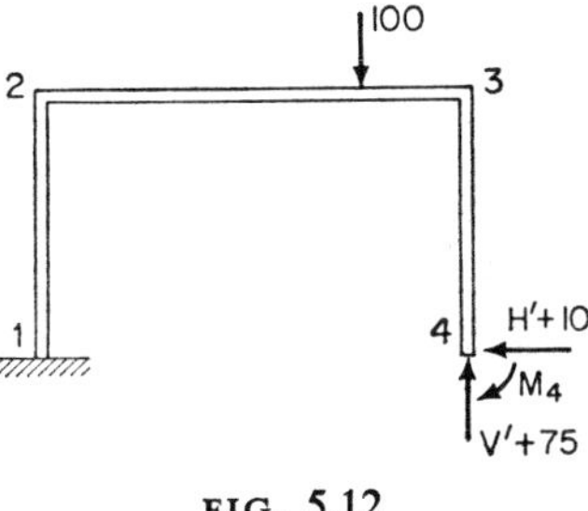

FIG. 5.12

$$\begin{bmatrix} 41{\cdot}76 & 42{\cdot}0 & -4{\cdot}02 \\ 42{\cdot}0 & 86{\cdot}7 & -5{\cdot}0 \\ -4{\cdot}02 & -5{\cdot}0 & 0{\cdot}47 \end{bmatrix} \begin{Bmatrix} H' \\ V' \\ M_4 \end{Bmatrix}$$

$$= \begin{Bmatrix} 2700 \\ 6188 \\ -337{\cdot}5 \end{Bmatrix} - \begin{bmatrix} 41{\cdot}76 & 42{\cdot}0 & -4{\cdot}02 \\ 42{\cdot}0 & 86{\cdot}7 & -5{\cdot}0 \\ -4{\cdot}02 & -5{\cdot}0 & 0{\cdot}47 \end{bmatrix} \begin{Bmatrix} -10 \\ +75 \\ 0 \end{Bmatrix}$$

$$= \begin{Bmatrix} -32{\cdot}4 \\ +105{\cdot}5 \\ -2{\cdot}7 \end{Bmatrix} \qquad (5.3)$$

Although the rather ill-conditioned flexibility matrix is unchanged, the values of the right-hand side vector of (5.3) are only a few per cent of those of (5.2) as the equation is now written in terms of the small corrective forces H' and V' rather than the actual reactions. If H' and V' are combined with the assumed reactions, then the solutions of equations (5.2) and (5.3) calculated to three significant figures are compared with each other and with a more accurate solution in the following table:

Reaction	Equation (5.2)	Equation (5.3)	Accurate solution
M_4	46·0	45·8	45·59
V	77·2	77·1	77·12
H	17·5	17·3	17·30

From this it can be seen that although the solution based on equation (5.2) is not at all bad, that based on (5.3) is significantly better with very little extra work involved in the calculations.

A discussion of the accuracy of solution and of the conditioning of sets of equations would be incomplete without the inclusion of one important point, which is, that if a set of equations in an analysis is ill conditioned, then the structure is particularly sensitive to changes in the particular redundant force or forces causing the ill-conditioning. In the arch of fig. 5.11, for example, the bending moment will be very considerably affected by small changes in the horizontal thrust. It is true that a different choice of unknowns will produce a more amenable set of equations; but this does not alter the physical fact that the structure is sensitive to small changes in the arch thrust. Care must therefore be taken in the design of indeterminate structures of this nature that full allowance is made for the effects of lack of fit in assembly or dimensional changes due to thermal expansion, otherwise extremely large internal forces may occur.

5.5 Symmetry and Antisymmetry

If a structure is symmetric about its centre-line, then any distribution of forces applied to it can be resolved into symmetric and antisymmetric (or skew symmetric) components. The structure can then be analysed for each component separately and the results superimposed. The forces applied to the structure of fig. 5.13(a), for example, may be resolved into the symmetric and antisymmetric force distributions of fig. 5.13(b and c). Further examples of symmetric and antisymmetric loadings are given in fig. 5.14.

When the forces applied to a structure are symmetric, the bending moment diagram and the displaced shape of the structure are also symmetric, but the shear force diagram is antisymmetric (fig. 5.15). At the centre-line of such a structure, therefore, the shear and the slope are both zero, which means that the right-hand half of the structure could be replaced by a set of rollers at the centre which allow no rotation and no shear force, as shown in fig. 5.15(d). Similarly, the bending moment distribution and the deflected shape of an antisymmetrically loaded structure are both antisymmetric and hence are zero at the centre-line, so that half the structure could be neglected if its members were taken to be pinned at the centre-line, as in fig. 5.15(h).

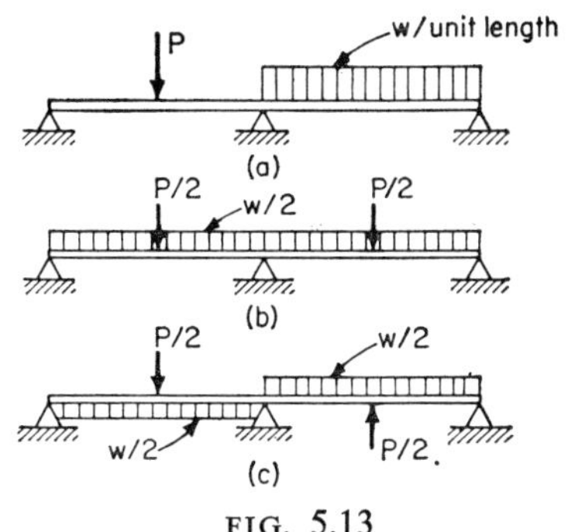

FIG. 5.13

The advantage of analysing a structure separately for the symmetric and antisymmetric components of its applied load is that the two parts each have fewer unknowns than the combined

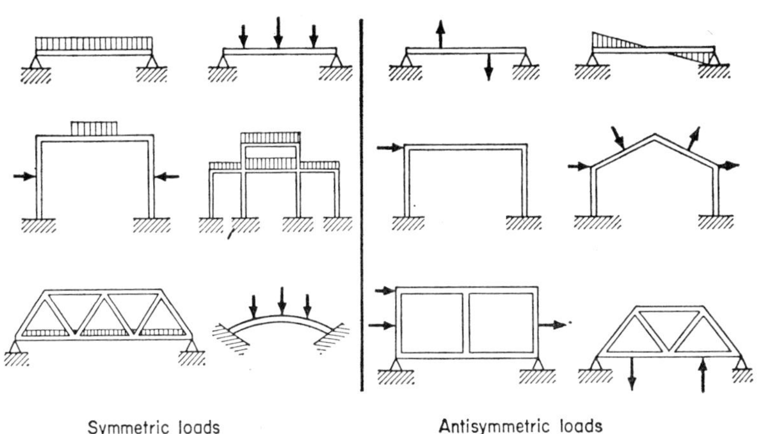

FIG. 5.14

structure. This is true both for a force and for a displacement analysis, but it will be illustrated here by a force method solution.

The portal frame example solved earlier in this chapter is not symmetric so that it cannot be handled by the symmetric/antisymmetric approach. If, however, all its members had the same section properties, then it could be solved as in fig. 5.16. The first step is to choose the unknowns. For the symmetric case, the moments at each of the four joints are taken as the redundants. Because of symmetry, however, we know that

$$M_2' = M_3' \quad \text{and} \quad M_1' = M_4'$$

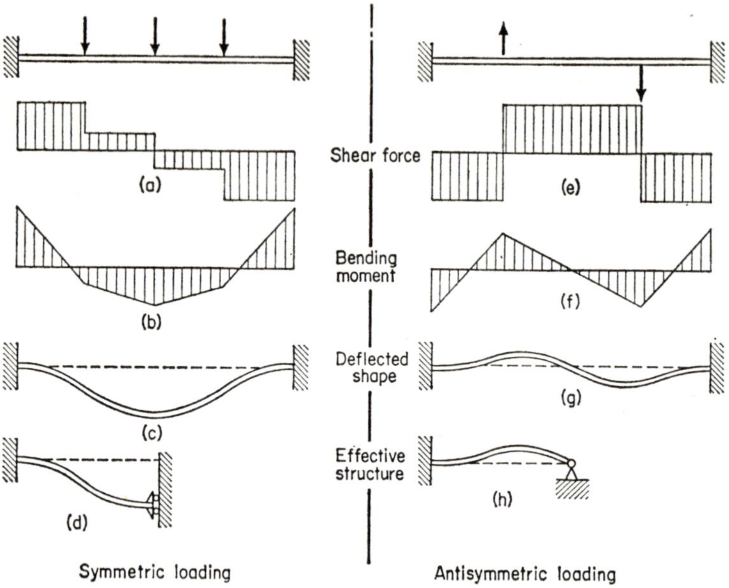

Shear force

(a)

(e)

Bending moment

(b)

(f)

Deflected shape

(c)

(g)

Effective structure

(d)

(h)

Symmetric loading

Antisymmetric loading

FIG. 5.15

so that there are only two independent unknowns. For the antisymmetric case, the moments at joints 2 and 3 are chosen, but as

$$M_2'' = -M_3''$$

only one unknown force needs to be found.

With the redundant forces set at zero, the structure becomes statically determinate. The four hinges inserted in the structure for the symmetric case do not make the structure a mechanism for it is subjected to the additional restraint of symmetry: this may be thought of as an external link applied to stop sideways movement of the structure. The two hinges inserted in the antisymmetric case at 2 and 3 appear to be insufficient to make the structure statically determinate, but it should be remembered that in effect there is a hinge in the middle of member 2–3, which brings the total number of releases in the structure to three.

First the applied loads and then the various unknown forces are applied to the statically determinate structures in fig. 5.16 (b, c and d), and the resulting rotations of the ends of the members are calculated. The sign convention used is that all clockwise rotations are positive. The compatibility equations are

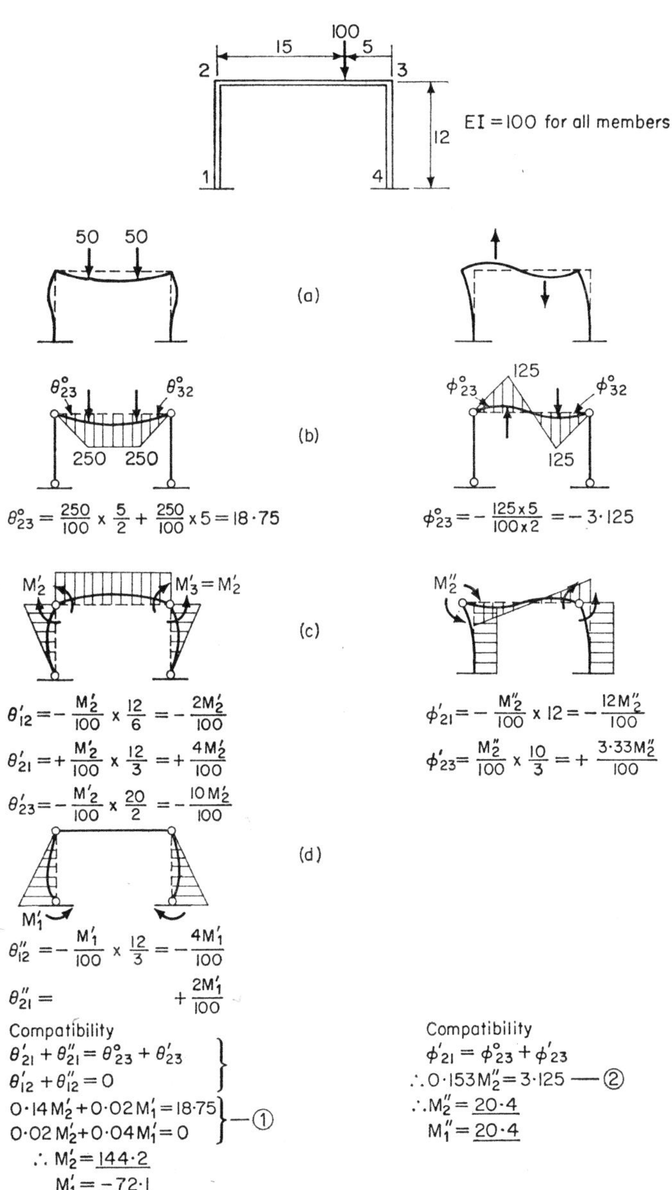

FIG. 5.16

then written; two for the symmetric case and one for the antisymmetric. These are solved to find the symmetric and antisymmetric bending moment distributions in the structure, which are plotted in the bending moment diagrams of fig. 5.17(a and b) and combined to give the final bending moment diagram in fig. 5.17(c).

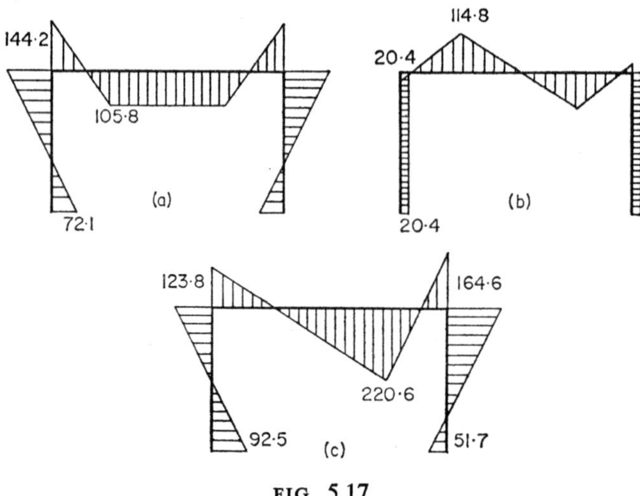

FIG. 5.17

It should be noticed that when a structure is solved by treating the symmetric and antisymmetric parts separately, the total number of unknowns is the same as if the structure were analysed as a whole. In this case the structure has a degree of indeterminacy of three, but two unknowns appear in the symmetric case while the other occurs in the antisymmetric case. Thus the effect of using the symmetry of the structure in this way is to uncouple one of the three compatibility equations from the other two. Writing equations (1) and (2) from fig. 5.16 in matrix form,

$$\begin{bmatrix} 0{\cdot}14 & 0{\cdot}02 & 0 \\ 0{\cdot}02 & 0{\cdot}04 & 0 \\ 0 & 0 & 0{\cdot}153 \end{bmatrix} \begin{Bmatrix} M_2' \\ M_1' \\ M_2'' \end{Bmatrix} = \begin{Bmatrix} 18{\cdot}75 \\ 0 \\ 3{\cdot}125 \end{Bmatrix} \tag{5.4}$$

The zero terms in the last row and column of the flexibility matrix indicate that M_2'' is not coupled to the other two unknowns and may be solved independently.

Besides making the compatibility equations easier to solve, the use of symmetry has other advantages, for by making use of it the total formulation of the problem is much simpler (the solution in this section may be compared with those in Sections 5.2 and 5.3) and the geometrical relationships may be handled more easily. Use should be made of the symmetry of a structure wherever possible.

5.6 Elastic Centre

The example of Section 5.3 analysed a portal frame by the force method choosing the moment and the vertical and horizontal reactions at one of the feet as the three unknowns. A flexibility matrix was obtained (equation (5.2)) relating the forces and displacements at the foot: none of the terms in the matrix was zero, so that a horizontal force, for example, would cause displacements in all three directions.

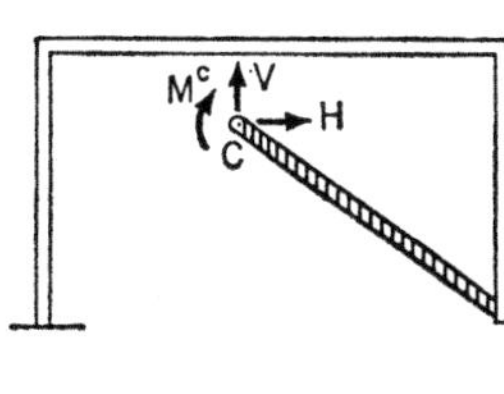

FIG. 5.18

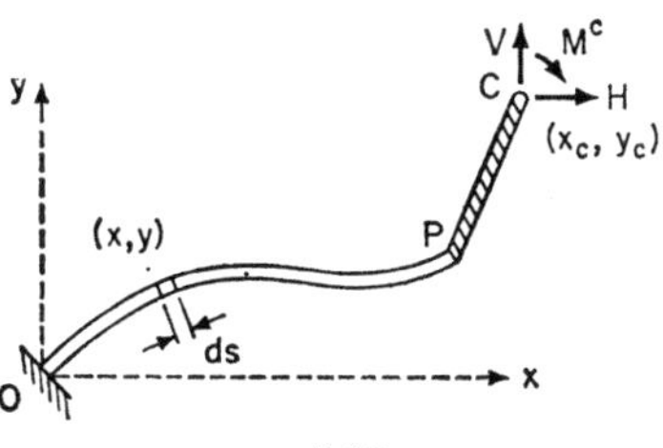

FIG. 5.19

If instead of applying the forces and the moment directly to the end of the frame we apply them to the end of a rigid arm which we imagine to be connected rigidly to the frame at its right-hand foot (fig. 5.18) then it is possible to find a suitable height of the end of the arm at which a horizontal force will cause horizontal and vertical displacements but no rotation. Similarly, a transverse position of the end of the arm can be found at which a vertical force will also cause no rotation. If the end of the arm fulfils these two criteria, its position is called the *elastic centre* of the frame. The advantage of choosing forces applied at the elastic centre for a force-method solution is that the flexibility matrix will be partly uncoupled (wholly uncoupled for a symmetric structure, as will be seen) so that the compatibility equations may be solved more easily.

To find the elastic centre for a general elastic chain (or singly connected) structure we may proceed as follows. Fig. 5.19 shows a structure OP of this type built-in at O, with a rigid arm PC attached to it at P. The problem is to find the position (x_C, y_C) of C such that the forces V and H will cause no rotation at C—which also implies that a moment applied at C will cause no vertical or horizontal displacement. If the moment at some point (x, y) on OP is $M(x, y)$, then the rotation between the ends of an element of length ds at that point is

$$d\theta = \frac{M(x, y)\,ds}{EI}$$

By geometry, the displacements at C due only to the flexure of the element ds at (x, y) are therefore

$$d\Delta_H = (y_C - y)\,d\theta = (y_C - y)\frac{M(x, y)\,ds}{EI}$$

$$d\Delta_V = -(x_C - x)\,d\theta = -(x_C - x)\frac{M(x, y)\,ds}{EI}$$

$$d\theta_C = d\theta = \frac{M(x, y)\,ds}{EI}$$

The bending moment at (x, y) is made up of a combination of the statically determinate bending moment $M^0(x, y)$ due to whatever applied loads act on the structure, and the effects of the redundant forces H, V and M^C applied at C, so that

$$M(x, y) = M^0(x, y) + H(y_C - y) - V(x_C - x) + M^C$$

Substituting this in the previous equations and integrating, we get

$$\left.\begin{aligned}
\Delta_H &= \int_0^p \frac{(y_C - y)M(x, y)\,ds}{EI} \\
&= \int_0^p \frac{(y_C - y)M^0(x, y)\,ds}{EI} + H\int_0^p \frac{(y_C - y)^2\,ds}{EI} \\
&\qquad - V\int_0^p \frac{(x_C - x)(y_C - y)\,ds}{EI} + M^C\int_0^p \frac{(y_C - y)\,ds}{EI} \\
\Delta_V &= -\int_0^p \frac{(x_C - x)M(x, y)\,ds}{EI} \\
&= -\int_0^p \frac{(x_C - x)M^0(x, y)\,ds}{EI} - H\int_0^p \frac{(x_C - x)(y_C - y)\,ds}{EI} \\
&\qquad + V\int_0^p \frac{(x_C - x)^2\,ds}{EI} - M^C\int_0^p \frac{(x_C - x)\,ds}{EI} \\
\theta_C &= \int_0^p \frac{M(x, y)\,ds}{EI} \\
&= \int_0^p \frac{M^0(x, y)\,ds}{EI} + H\int_0^p \frac{(y_C - y)\,ds}{EI} \\
&\qquad - V\int_0^p \frac{(x_C - x)\,ds}{EI} + M^C\int_0^p \frac{ds}{EI}
\end{aligned}\right\} \quad (5.5)$$

So far, these equations are perfectly general. We now want to find x_C and y_C such that a horizontal and a vertical force will cause no rotation. For this to be so, the coefficients of H and V in the expression for θ_C must be zero. Thus

$$\int_0^p \frac{(y_C - y)\,ds}{EI} = 0; \qquad \int_0^p \frac{(x_C - x)\,ds}{EI} = 0$$

Hence, as y_C and x_C are constants, the position of the elastic centre is given by

$$y_C = \frac{\displaystyle\int_0^p \frac{y\,ds}{EI}}{\displaystyle\int_0^p \frac{ds}{EI}}; \qquad x_C = \frac{\displaystyle\int_0^p \frac{x\,ds}{EI}}{\displaystyle\int_0^p \frac{ds}{EI}} \tag{5.6}$$

The elastic centre is thus at the centre of gravity of the structure weighted with $1/EI$.

To simplify some of the subsequent expressions, the origin is now changed to the elastic centre C. If X and Y are the new coordinates, the necessary transformation is

$$x = X + x_C$$
$$y = Y + y_C$$

Substituting these into equations (5.5), and rewriting them in matrix form, we get

$$\begin{bmatrix} \displaystyle\int_0^p \frac{Y^2\,ds}{EI} & \displaystyle -\int_0^p \frac{XY\,ds}{EI} & 0 \\ \displaystyle -\int_0^p \frac{XY\,ds}{EI} & \displaystyle\int_0^p \frac{X^2\,ds}{EI} & 0 \\ 0 & 0 & \displaystyle\int_0^p \frac{ds}{EI} \end{bmatrix} \begin{Bmatrix} H \\ V \\ M^C \end{Bmatrix} = \begin{Bmatrix} \displaystyle\int_0^p \frac{M^0(X, Y)\,Y\,ds}{EI} \\ \displaystyle -\int_0^p \frac{M^0(X, Y)X\,ds}{EI} \\ \displaystyle -\int_0^p \frac{M^0(X, Y)\,ds}{EI} \end{Bmatrix} \tag{5.7}$$

The first four terms in the flexibility matrix on the left-hand side may be thought of as the moments of inertia and product of inertia of the structure inversely weighted with the section stiffness EI, and the last term, $\int_0^p ds/EI$, represents the weighted 'area' of the structure. The off-diagonal terms will evidently all be zero if the coordinates are the principal axes of the moment of inertia of the structure. It would never in practice be worth while to align the coordinate directions with the principal axes, but if the structure is symmetric the coordinates will automatically lie along the principal axes. In such a case there would be no coupling whatsoever between the unknowns and each could be solved independently.

The first two elements of the vector on the right-hand side of equation

(5.7) are the first moments of the M^0/EI distribution taken about the elastic centre, and the last term is its 'area'.

The original example of fig. 5.1 will now be solved. First, the position of the elastic centre must be found. The weighted area of the structure is

$$\int_1^4 \frac{ds}{EI} = \int_1^2 \frac{dy}{80} + \int_2^3 \frac{dx}{100} + \int_3^4 \frac{dy}{100}$$

$$= \frac{12}{80} + \frac{20}{100} + \frac{12}{100} = 0{\cdot}47$$

The position of the elastic centre is then given by, from equation (5.6),

$$x_C = \left(\frac{20}{100} \times 10 + \frac{12}{100} \times 20\right) \frac{1}{0{\cdot}47} = 9{\cdot}36$$

$$y_C = \left(\frac{12}{80} \times 6 + \frac{20}{100} \times 12 + \frac{12}{100} \times 6\right) \frac{1}{0{\cdot}47} = 8{\cdot}56$$

The 'moment of inertia' terms in the flexibility matrix are now found by finding the moments of inertia of each member about its centre of gravity and then using the parallel axis theorems.

$$\int_1^4 Y^2\,ds = \frac{12^3}{12 \times 80} + \frac{12^3}{12 \times 100} + \left(\frac{12}{80} + \frac{12}{100}\right) \times 2{\cdot}56^2 + \frac{20}{100} \times 3{\cdot}44^2 = 7{\cdot}37$$

$$\int_1^4 X^2\,ds = \frac{20^3}{12 \times 100} + \frac{12}{80} \times 9{\cdot}36^2 + \frac{12}{100} \times 10{\cdot}64^2 + \frac{20}{100} \times 0{\cdot}64^2 = 33{\cdot}5$$

$$\int_1^4 XY\,ds = \frac{12}{80} \times (-9{\cdot}36) \times (-2{\cdot}56) + \frac{20}{100} \times 0{\cdot}64 \times 3{\cdot}44 + \frac{12}{100} \times 10{\cdot}64 \times (-2{\cdot}56) = 0{\cdot}77$$

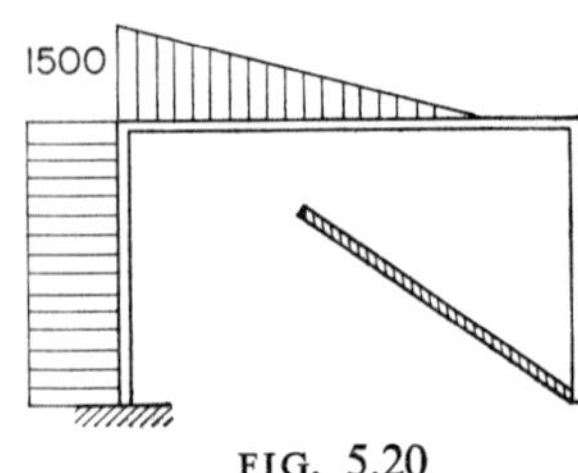

FIG. 5.20

Fig. 5.20 shows the statically determinate bending moment diagram due to the applied force. Using this bending moment distribution, the applied load terms on the right-hand side of equation (5.7) become

$$-\Delta_H^0 = \int_1^4 \frac{M^0 Y\,\mathrm{d}s}{EI} = \frac{1500 \times 12}{80} \times (-2{\cdot}56) = -190$$

$$-\Delta_V^0 = -\int_1^4 \frac{M^0 X\,\mathrm{d}s}{EI} = -\frac{1500 \times 12}{80} \times (-9{\cdot}36) - \frac{1500 \times 15}{2 \times 100} \times (-4{\cdot}36)$$
$$= +2595$$

$$-\theta_C^0 = -\int_1^4 \frac{M^0\,\mathrm{d}s}{EI} = -\frac{1500 \times 12}{80} - \frac{1500 \times 15}{2 \times 100} = -337{\cdot}5$$

Hence

$$\begin{bmatrix} 7{\cdot}37 & -0{\cdot}77 & 0 \\ -0{\cdot}77 & 33{\cdot}5 & 0 \\ 0 & 0 & 0{\cdot}47 \end{bmatrix} \begin{Bmatrix} H \\ V \\ M^C \end{Bmatrix} = \begin{Bmatrix} -190 \\ +2595 \\ -337{\cdot}5 \end{Bmatrix} \qquad (5.8)$$

The off-diagonal term is very small, so the equations are well conditioned and an accurate solution can be obtained. The solution is

$$H = -17{\cdot}7$$
$$V = 77{\cdot}1$$
$$M^C = -718$$

From these results the various joint moments may be found by statics to be

$$M_1 = 91 \qquad M_3 = 168{\cdot}5$$
$$M_2 = 124 \qquad M_4 = 47{\cdot}5$$

but some accuracy is lost in obtaining these values as small differences between large numbers are involved in their calculation: M^C is very much larger than the joint moments and so are the moments of H and V taken about the joints. This loss of accuracy is not altogether surprising in view of the discussion in Section 5.4 in which it was stated that an important factor in the accuracy of a force-method solution was the closeness with which the statically determinate system approximated to the behaviour of the actual structure.

The elastic centre approach is the basis of the *column-analogy method* which uses the analogy between the elastic centre equations and the stress distribution in a short column subjected to an eccentric axial force to develop a more formalised computational procedure. Because the column-analogy method automatically takes care of all signs, it has some advantages; but as it is necessary to remember which terms are analogous to each other, it is hardly worth while learning the method unless many similar calculations have to be carried out. It will not be dealt within this book.

PROBLEMS

5.1 A beam ABCD of length 60 spans an opening and is simply supported at its ends. It rests on the centres of two transverse beams at its third points B and C: the transverse beams both have a length of 40 and are simply supported at their ends.

Use the force method to find the deflection at B due to a concentrated load of 15 applied at the centre of beam ABCD. Neglect torsional effects. All beams have a uniform section stiffness of $EI = 2 \times 10^8$.

Answer: 0·0828

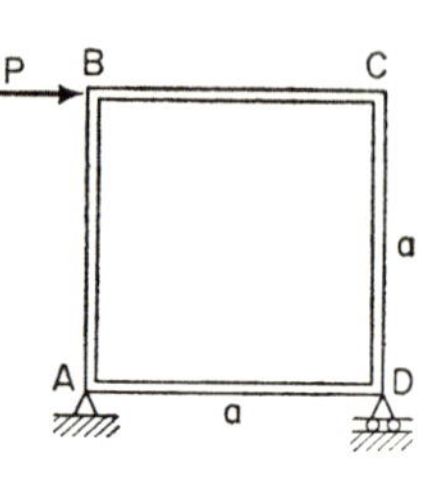

FIG. 5.21

5.2 The uniform square ring ABCD of fig. 5.21 is rigidly jointed and is attached to supports by pins at A and D. If a force P is applied horizontally at B, find the bending moment at A in terms of P and the side length a.

Answer: $Pa/4$

5.3 The portal frame shown in fig. 5.22 is built-in at point 1 and hinged at point 4. The section stiffness is the same for all members. Use the force method to find the three reactions at point 1 by
(a) choosing the two reactions at point 4 as unknowns:
(b) choosing the bending moments at points 2 and 3 as unknowns.
Also find the position of the elastic centre.

Answer: 4·85, 0·839, −3·2

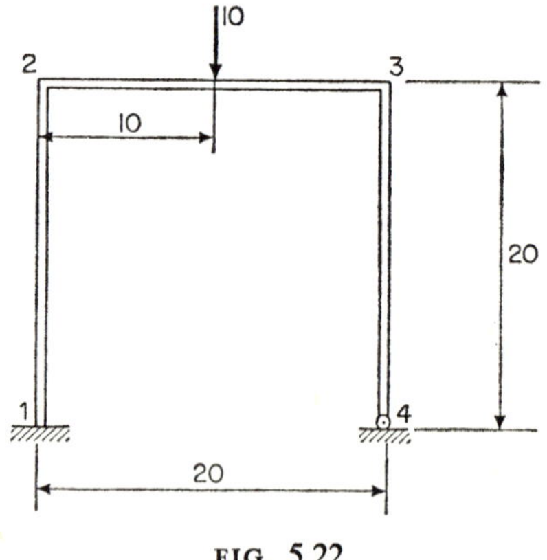

FIG. 5.22

5.4 Use the elastic centre method to find the joint bending moments in the frame shown in fig. 5.23, and sketch the bending moment diagram. All members are uniform, and the value of the section stiffness EI of the two vertical members is twice that of the horizontal member.

Answer: 2800, 1200

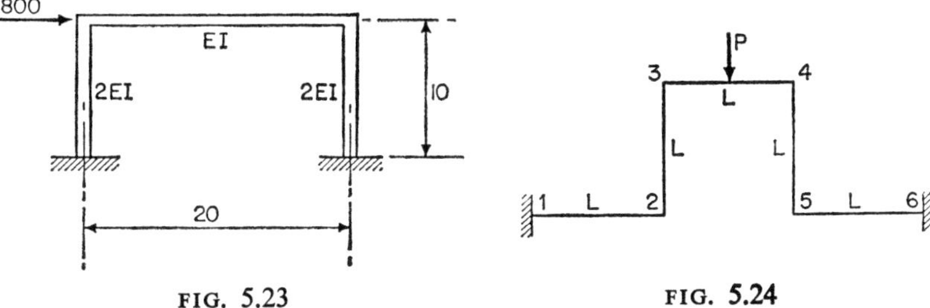

FIG. 5.23

FIG. 5.24

5.5 Use the elastic centre method to find the moments in terms of P and L at points 1 and 6 of the symmetrically loaded frame shown in fig. 5.24. All members are uniform and have the same sectional properties.

Answer: $0{\cdot}298PL$

5.6 A uniform square frame similar to that shown in fig. 5.21 has sides of length 10 units and is subjected to a horizontal force of 2 applied at the mid-point of member AB. Find the joint bending moments and sketch the bending moment diagram. Use the elastic centre method.

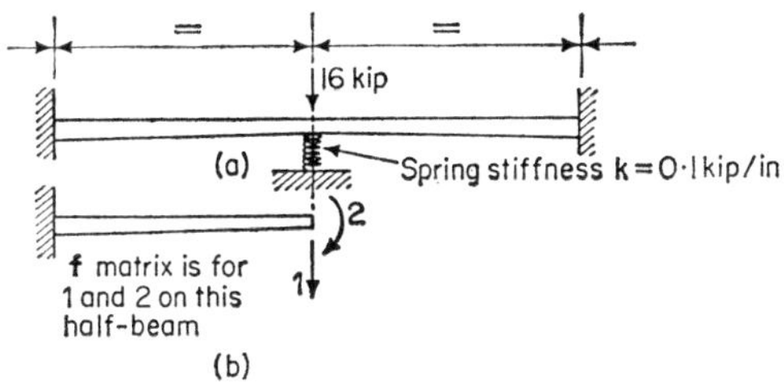

FIG. 5.25

5.7 Use the force method to find (a) the central moment in the beam shown in fig. 5.25(a), and (b) the force in the spring. The beam is symmetrical and the flexibility matrix for one half of the beam acting as a cantilever (fig. 5.25(b)) is

$$\mathbf{f} = \begin{bmatrix} 40 & 5 \\ 5 & 1 \end{bmatrix}$$

6

Displacement Methods of Frame Analysis

6.1 Introduction

The displacement method has already been outlined in general terms in Chapter 2, where its complementary relationship with the force method has been discussed. The present chapter introduces formalised analysis procedures based on the general principles of the method, and applies them to various types of frame structure. It is essential to note that all the following work adheres rigidly to the sign convention first introduced on page 52, that

(a) clockwise moments applied to the ends of beams are positive
(b) clockwise joint rotations are positive
(c) clockwise chord rotations are positive.

6.2 Slope Deflection

The *slope-deflection* method was developed by Axel Bendixen in 1914. It was the most frequently used method of frame analysis until Hardy Cross introduced the technique of *moment distribution* in 1932 which will be discussed in a subsequent section. There are still a few situations in which slope deflection is a better method than moment distribution, and interest in it has revived in recent years as it has become the basis of many structural analysis computer programs.

Slope-deflection equations formally express the moments acting on the ends of a beam in terms of its end rotations and displacements. Fig. 6.1(a) shows a typical beam element in a structure connecting joints i and j. The joints have been given rotations θ_i and θ_j, while their displacement relative to one another is expressed by the chord rotation φ_{ij}. The combined action of these displacements results in end moments M_{ij} and M_{ji} being applied to the beam. The end moments are found separately for each displacement in turn: figs. 6.1 (b and c) shows the effects of the two end rotations θ_i and θ_j. The moments they cause are expressed in terms of the stiffness and carry-over factors defined in Section 3.3, while if the chord rotation is treated as a combination of the two separate end rotations φ_{ij} as shown in figs. 6.1 (d and e), the moments due to it may also be written in terms of stiffness and carry-over factors. In the figure, it should be noted that all moments are drawn clockwise, following the sign convention adopted. This is not meant to imply that the *actual* moment is positive in all cases: it is evident that a positive chord rotation will lead to negative moments, but as long as the clockwise positive convention is strictly followed, the signs will automatically be correct.

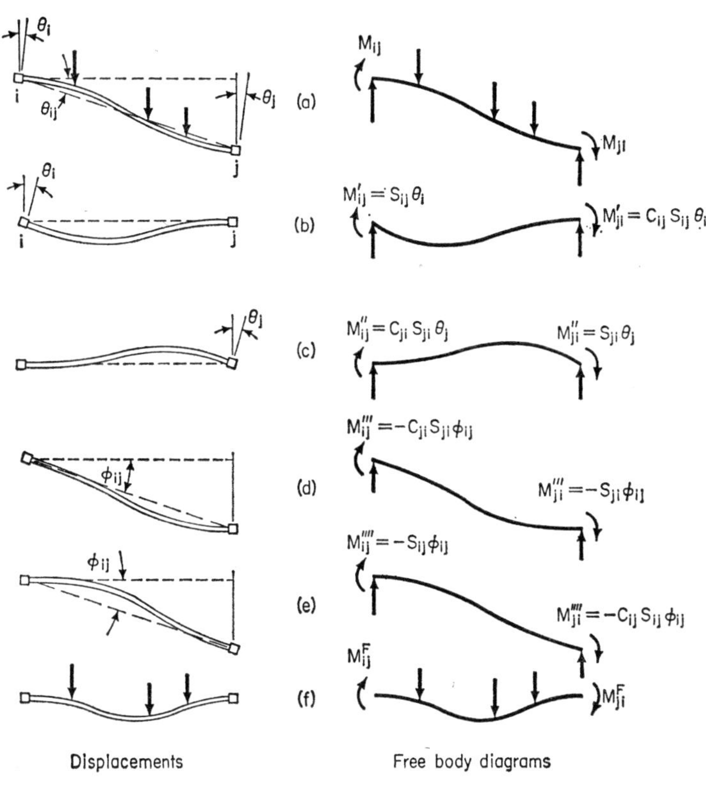

FIG. 6.1

Finally, the applied load acting on the beam will cause fixed-end moments to act on the ends of the beam in addition to those caused by the displacements. Summing the five components of the total moment we get

$$M_{ij} = M^F_{ij} + S_{ij}(\theta_i - \varphi_{ij}) + C_{ji}S_{ji}(\theta_j - \varphi_{ij}) \tag{6.1}$$

This is the general form of the slope-deflection equation. It holds for any beam, prismatic or otherwise, and also for any end conditions, provided the correct modified stiffness factors are used as, for example, those given in Table 3.1.

The slope deflection method then proceeds as follows:

(a) With all joints fixed against rotation and displacement, fixed-end moments are calculated for loaded members.
(b) Stiffness and carry-over factors are found for all members.
(c) Slope-deflection equations are written down expressing beam-end moments in terms of displacements.
(d) The requisite number of equilibrium equations is written down in terms of moments. There will be as many equilibrium equations as unknown displacements.
(e) Using the slope-deflection equations, the equilibrium equations are written in terms of displacements and are then solved to find the displacements.
(f) The beam-end moments are found by back-substituting the displacements into the slope-deflection equations.

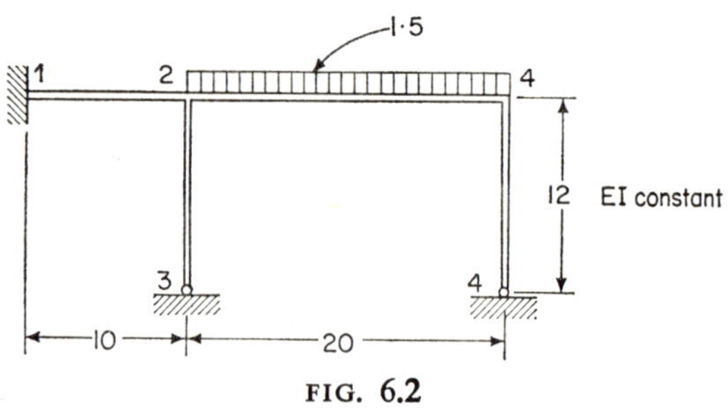

FIG. 6.2

As an example of the method, consider the frame shown in fig. 6.2, all of whose members are uniform and have the same section stiffness of, say, $EI = 10$. Because of the geometry of the structure neither joint can displace laterally, so that no member will have a chord rotation. The unknown displacements which must be found are therefore the two joint rotations θ_2 and θ_4. We now follow the scheme laid out above.

(a) Fixed-end moments

$$M_{24}^F = -\frac{wL^2}{12} = -\frac{1{\cdot}5 \times 20^2}{12} = -50$$

$$M_{42}^F = +50$$

and all other fixed-end moments are zero.

(b) Stiffness and carry-over factors. Using the results of Table 3.1 we have, taking $EI = 10$,

$$S_{21} = \frac{4 \times 10}{10} = 4 \qquad S'_{32} = \frac{3 \times 10}{12} = 2{\cdot}5$$

$$S_{24} = \frac{4 \times 10}{20} = 2 \qquad S'_{45} = \frac{3 \times 10}{12} = 2{\cdot}5$$

$$S_{42} = \frac{4 \times 10}{20} = 2$$

Notice that S'_{23} and S'_{45} have been modified to allow for the pin joints at 3 and 4. All the carry-over factors are $\frac{1}{2}$ except for C_{23} and C_{45} which are zero.

(c) Slope-deflection equations. From equation (6.1) we have

$$M_{21} = M^F_{21} + S_{21}(\theta_2 - \varphi_{12}) + C_{12}S_{12}(\theta_1 - \varphi_{12})$$

Remembering that M^F_{21}, θ_1 and φ_{12} are all zero and substituting the correct values of stiffness and carry-over factors, the equation becomes simply

$$M_{21} = 4\theta_2$$

Substituting directly into equation (6.1) the other moments become

$$\begin{aligned} M_{12} &= \tfrac{1}{2} \times 4\theta_2 = 2\theta_2 \\ M_{23} &= 2{\cdot}5\theta_2 \\ M_{24} &= -50 + 2\theta_2 + \tfrac{1}{2} \times 2\theta_4 \\ M_{42} &= 50 + 2\theta_4 + \tfrac{1}{2} \times 2\theta_2 \\ M_{45} &= 2{\cdot}5\theta_4 \end{aligned}$$

(d) Equilibrium equations. There are two unknowns so two equilibrium equations must be chosen. Out of the infinite number which could be formed for the structure, the obvious choices are the equations of rotational equilibrium of the joints,

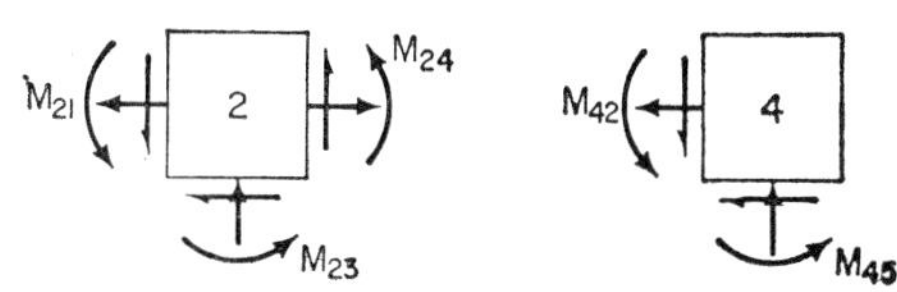

FIG. 6.3 Free body diagrams of joints

for no direct forces come into these two equations and they are expressed only in terms of beam-end moments. If a clockwise moment is applied to the end of a beam by a joint, then conversely the beam must apply an anticlockwise moment to the joint. The moments acting on joints 2 and 4 are shown in fig. 6.3. The total moment acting on a joint must be zero, so assuming that the joints are small and that shear forces will cause no appreciable moment, the equilibrium equations become

$$M_{21} + M_{23} + M_{24} = 0$$
$$M_{42} + M_{45} = 0$$

(e) Solution for displacements. Substituting the values of the moments given by the slope-deflection equations into the equilibrium equations, they become

$$4\theta_2 + 2{\cdot}5\theta_2 - 50 + 2\theta_2 + \theta_4 = 0$$
$$50 + 2\theta_4 + \theta_2 + 2{\cdot}5\theta_4 = 0$$

Rearranging terms,

$$8{\cdot}5\theta_2 + \theta_4 = 50$$
$$\theta_2 + 4{\cdot}5\theta_4 = -50$$

which may be written in matrix form as

$$\begin{bmatrix} 8{\cdot}5 & 1 \\ 1 & 4{\cdot}5 \end{bmatrix} \begin{Bmatrix} \theta_2 \\ \theta_4 \end{Bmatrix} = \begin{Bmatrix} 50 \\ -50 \end{Bmatrix}$$

The square matrix is the stiffness matrix for joint rotations, and the vector on the right-hand side represents the moments applied to joints 2 and 4 by the beam 2–4 when acted on by the applied load. These moments may be regarded as a statically equivalent load—a point discussed at the beginning of Section 3.4.

Solving the equations gives

$$\theta_2 = +7{\cdot}38$$
$$\theta_4 = -12{\cdot}77$$

It should be noticed that the stiffness matrix is strongly diagonal so that the equations are well conditioned and their solution will be accurate. This well-conditioning is generally true of displacement-method analyses and is one of the advantages of the method over force-method analyses which, as has been seen in the previous chapter, can often lead to ill-conditioned equations.

(f) Back-substitution. The moments can now be found by substituting the joint rotations into the slope-deflection equations to give

$$M_{12} = 2 \times 7{\cdot}38 = +14{\cdot}76$$
$$M_{21} = 4 \times 7{\cdot}38 = +29{\cdot}52$$
$$M_{24} = 2 \times 7{\cdot}38 - 12{\cdot}77 - 50 = -48{\cdot}01$$
$$M_{23} = 2{\cdot}5 \times 7{\cdot}38 = 18{\cdot}45$$
$$M_{42} = -2 \times 12{\cdot}77 + 7{\cdot}38 + 50 = 31{\cdot}84$$
$$M_{45} = -2{\cdot}5 \times 12{\cdot}77 = -31{\cdot}92$$

These moments are used to draw the bending moment diagram of fig. 6.4. Joint 2 is less flexible than joint 4 due to the stiffening effect of member 2–1. It can be seen that the stiffer joint attracts more moment

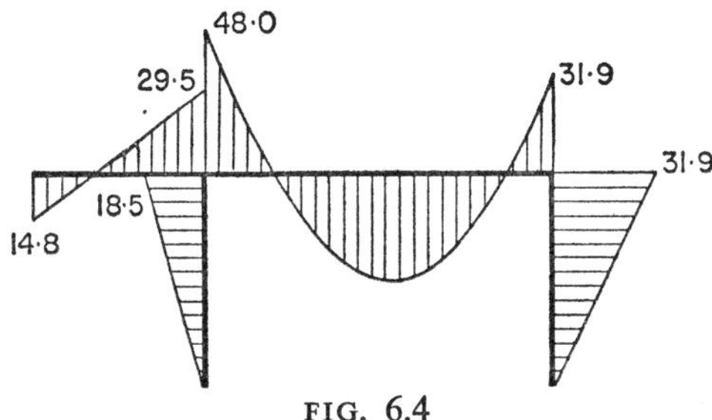

FIG. 6.4

than the other. This illustrates a general principle which is important in design: that the stiffer one part of a structure is made relative to the rest, the more load it will tend to attract to itself.

6.3 Sway

The only displacement unknowns in the previous example were joint rotations: translational displacements could not take place in any part of the structure. In such a case the formulation of the problem is simple as the rotation of each joint is independent of the rotations of all the others. However (if axial deformation of members is neglected), nodes will not in general translate independently of each other.

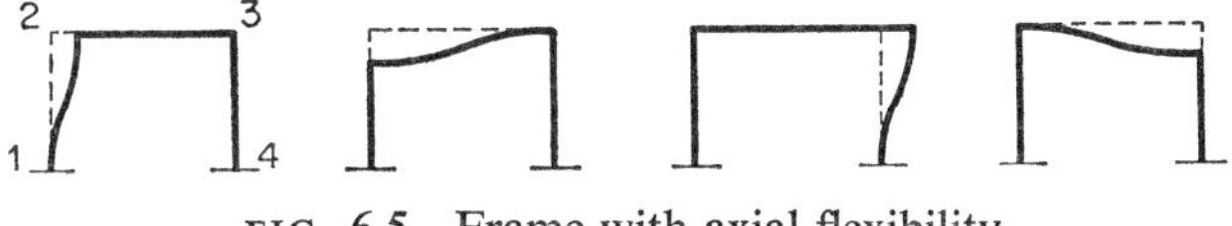

FIG. 6.5 Frame with axial flexibility

For example, assuming the portal frame of fig. 6.5 does have significant axial flexibility in its members, then we could take the four displacements shown in the diagram as translational unknowns in an analysis; but with an axially rigid frame points 2 and 3 cannot move vertically and can only move transversely by the same amount as shown in fig. 6.6. Thus the frame is subjected to the restricting relationships

$$\varphi_{23} = 0$$
$$\varphi_{12} = \varphi_{34}$$

between the chord rotations of its members.

One of the first things to do in the analysis of a frame with translational displacements (often called *sidesway*) is to evaluate the chord rotation relationships. Fig. 6.7 shows a frame to which a sway displacement has been applied. It is important to realise that the deflected shape (b) is not at this

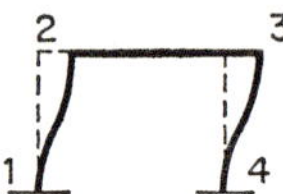

FIG. 6.6 Axially rigid frame

stage the actual displaced shape of the structure but only the effect of the sway displacement: the other two possible nodal displacements of the structure—the joint rotations—are not applied. The horizontal displacements of nodes 2 and 3 are the same, that is

$$\Delta_2 = \Delta_3$$

and the chord rotation relationship is therefore

$$20\varphi_{12} = 30\varphi_{34}$$

This is a purely geometrical relationship. It does not matter if the various members have different section stiffnesses, and neither does it make any

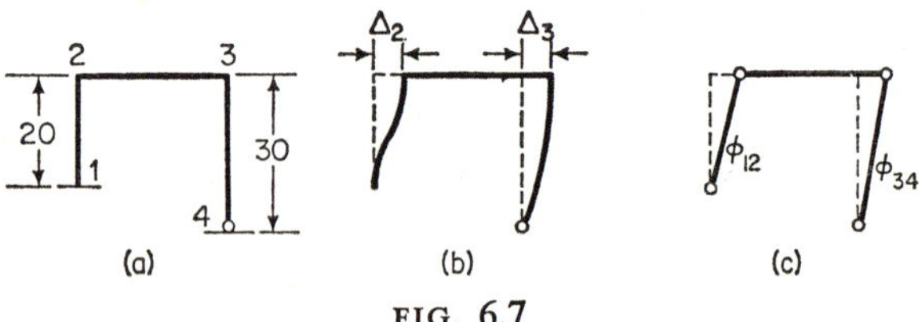

FIG. 6.7

difference that joint 4 happens to be pinned, not fixed. In order to find the kinematic relationship between the chord rotations more easily, we would therefore be justified in replacing the actual frame by the imaginary mechanism of fig. 6.7(c) with hinges at all the nodes. This mechanism will be called the *equivalent sway mechanism* of the frame. A further example of a frame and its sway mechanism is given in fig. 6.8, from which the relationships between the chord rotations can be seen to be

$$\varphi_{23} = -\varphi_{12}$$
$$2\varphi_{45} = \varphi_{12}$$

and so on. The concept of the sway mechanism will be discussed more fully in Section 6.9, but it should be borne in mind that the structure itself does not actually behave as a mechanism: the only use for the sway mechanism is as a help in evaluating the geometry of the displacement of the structure.

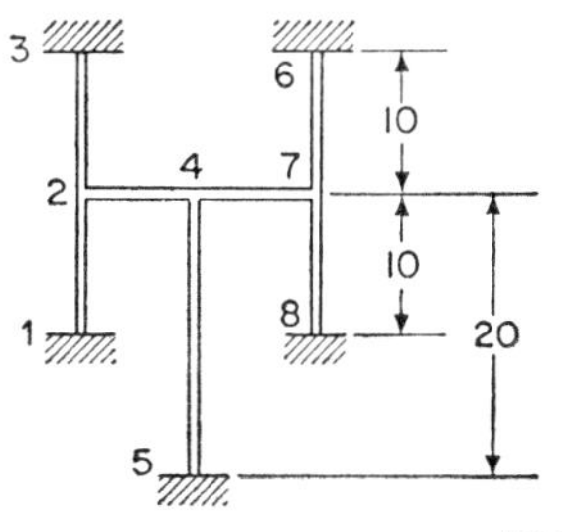

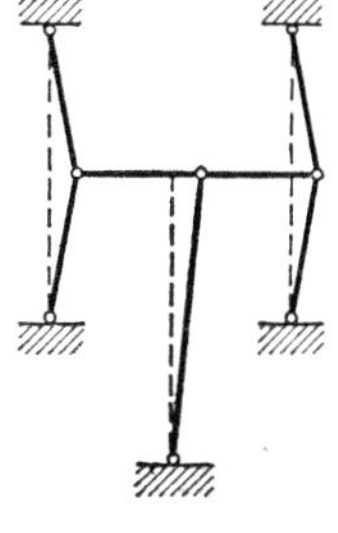

FIG. 6.8

The frame solved in the previous chapter by the force method is now analysed using the slope-deflection approach. The frame itself is redrawn in fig. 6.9, together with its equivalent mechanism. There are evidently three unknown displacements in the problem: θ_2, θ_3 and φ_{21}. Following the pattern laid down in the previous section, the calculation proceeds as follows:

(a) Fixed-end moments. From equation (3.14),

$$M^F_{23} = -\frac{100 \times 5^2 \times 15}{20^2} = -93{\cdot}8$$

$$M^F_{23} = +\frac{100 \times 15^2 \times 5}{20^2} = +281$$

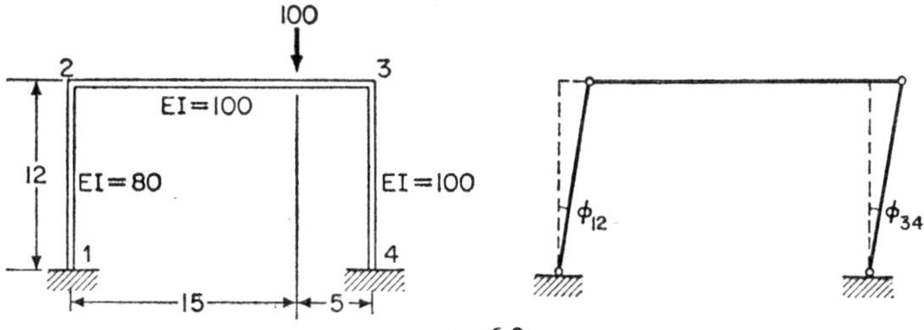

FIG. 6.9

(b) Stiffness and carry-over factors

$$S_{21} = \frac{4 \times 80}{12} = 26{\cdot}7 \qquad S_{23} = \frac{4 \times 100}{20} = 20$$

$$S_{32} = \frac{4 \times 100}{20} = 20 \qquad S_{34} = \frac{4 \times 100}{12} = 33{\cdot}3$$

All carry-over factors are $\frac{1}{2}$.

(c) Slope-deflection equations. From equation (6.1),

$$M_{12} = M_{12}^F + S_{12}(\theta_1 - \varphi_{12}) + C_{21}S_{21}(\theta_2 - \varphi_{12})$$

which becomes, as $M_{12}^F = 0$ and $\theta_1 = 0$,

$$\begin{aligned} M_{12} &= C_{21}S_{21}\theta_2 - (S_{12} + C_{21}S_{21})\varphi_{12} \\ &= 13{\cdot}3\theta_2 - (26{\cdot}7 + \tfrac{1}{2} \times 26{\cdot}7)\varphi_{12} \end{aligned}$$

Similarly

$$\begin{aligned} M_{21} &= 26{\cdot}7\theta_2 - (26{\cdot}7 + \tfrac{1}{2} \times 26{\cdot}7)\varphi_{12} \\ M_{23} &= -93{\cdot}8 + 20\theta_2 + 10\theta_3 \\ M_{32} &= 281 + 20\theta_3 + 10\theta_2 \\ M_{34} &= 33{\cdot}3\theta_3 - (33{\cdot}3 + \tfrac{1}{2} \times 33{\cdot}3)\varphi_{12} \\ M_{43} &= 16{\cdot}7\theta_3 - (33{\cdot}3 + \tfrac{1}{2} \times 33{\cdot}3)\varphi_{12} \end{aligned}$$

(d) Equilibrium equations. Moment equilibrium about nodes 1 and 2 gives the equations

$$\begin{aligned} M_{21} + M_{23} &= 0 \\ M_{32} + M_{34} &= 0 \end{aligned}$$

There are three unknowns in the problem, so one other equilibrium equation is needed. It must be expressed in terms of moments alone because of the nature of the slope-deflection equations. Such an equation could be found in this way: consider members 1–2 and 3–5 (fig. 6.10). Then taking moments about 2 and 3,

$$\begin{aligned} M_{12} + M_{21} + 12H &= 0 \\ M_{43} + M_{34} - 12H &= 0 \end{aligned}$$

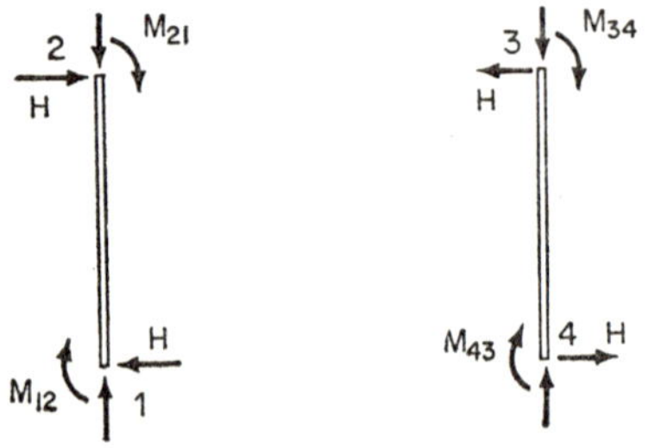

Eliminating H from these equations gives

$$M_{12} + M_{21} + M_{43} + M_{34} = 0$$

FIG. 6.10

which is an equilibrium equation of the required form. Such an equation will be called a *moment-sway equation*. The same equation may be found more elegantly and rapidly, however, by the use of virtual work. Consider the structure as a series of separate free bodies as shown in fig. 6.11(a), and give the total set of members a virtual displacement as in (b). When setting up an equation of equilibrium using virtual work we have of course an unlimited choice of possible displacements or combinations of displacements, but the solution becomes particularly simple if the displacement chosen has the same shape as that chosen for the sway mechanism, for in such a case the work done by all linear forces is zero and only moments enter into the equation. For equilibrium the virtual work is zero and

$$M_{12}\,\delta\varphi_{12} + M_{21}\,\delta\varphi_{12} + M_{34}\,\delta\varphi_{34} + M_{43}\,\delta\varphi_{34} = 0$$

or, as $\delta\varphi_{12} = \delta\varphi_{34}$,

$$M_{12} + M_{21} + M_{34} + M_{43} = 0$$

as before. With a little experience of the virtual work formulation, this equation could have been written straight down. Note that in this example the applied load is displaced perpendicular to its line of action so that it does no work and does not enter the equilibrium equation.

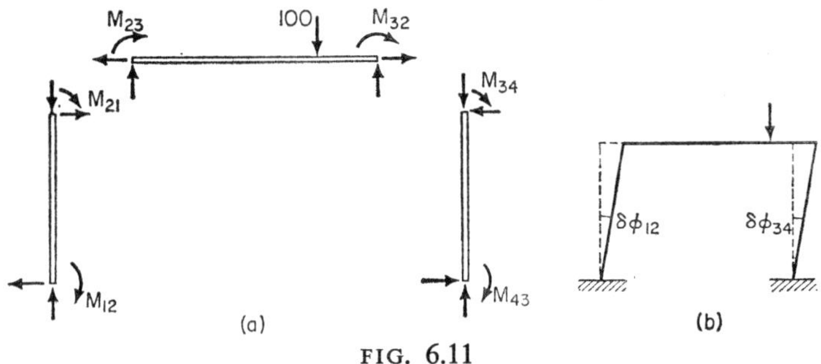

FIG. 6.11

(e) *Solution of equations.* Expanding the three equilibrium equations in terms of displacements by using the slope-deflection equations, we obtain

$$26{\cdot}7\theta_2 - 40\varphi_{12} - 93{\cdot}8 + 20\theta_2 + 10\theta_3 = 0$$

$$281 + 20\theta_3 + 10\theta_2 + 33{\cdot}3\theta_3 - 50\varphi_{12} = 0$$

$$13{\cdot}3\theta_2 - 40\varphi_{12} + 26{\cdot}7\theta_2 - 40\varphi_{12} + 16{\cdot}7\theta_3 - 50\varphi_{12} + 33{\cdot}3\theta_3 - 50\varphi_{12} = 0$$

Rearranging terms and writing the equations in matrix form gives

$$\begin{bmatrix} 46{\cdot}7 & 10 & -40 \\ 10 & 53{\cdot}3 & -50 \\ -40 & -50 & 180 \end{bmatrix} \begin{Bmatrix} \theta_2 \\ \theta_3 \\ \varphi_{12} \end{Bmatrix} = \begin{Bmatrix} 93{\cdot}8 \\ -281 \\ 0 \end{Bmatrix}$$

The first two diagonal terms of the stiffness matrix are of the same order as the off-diagonal terms in the last row and column so that the matrix is not as well conditioned as in the previous slope-deflection example.

Solving the equations gives

$$\begin{aligned} \phi_{12} &= -1{\cdot}469 \\ \theta_3 &= -7{\cdot}06 \\ \theta_2 &= +2{\cdot}25 \end{aligned}$$

(f) Back-substitution. The frame moments are obtained by substituting the displacements into the slope-deflection equations. This gives

$$\begin{aligned} M_{12} &= 88{\cdot}7 \\ M_{21} &= 118{\cdot}6 \\ M_{34} &= -161{\cdot}5 \\ M_{43} &= -44{\cdot}0 \end{aligned}$$

which agree with the results previously found by the force method.

6.4 Moment Distribution

For any but the simplest of structures, the slope-deflection method suffers from the disadvantage that a large number of simultaneous equations must be solved. This difficulty was largely overcome when Hardy Cross invented the *moment-distribution* method in 1932. In many ways moment distribution is very similar to slope deflection though this similarity is largely disguised by fundamental differences in the technique of solution. Both methods use the same carry-over factors, stiffness factors and fixed-end moments, and both use the same approach to formulating the kinematic relations of sidesway. However, whereas the slope-deflection approach uses both joint-equilibrium and sway-equilibrium equations, moment distribution only formulates the latter explicitly while the joint-equilibrium equations are solved by an iterative process in terms not of displacements (though moment distribution is a displacement method) but of moments. Yet moment distribution is more than a convenient method of successive approximations for the solution of a set of equations, for it is a process whose physical interpretation can be of considerable help in understanding the structural action of the frame being analysed.

Moment distribution is best described with reference to an actual example and to this end the frame shown in fig. 6.12 will be analysed.

First, imagine that the structure cannot sway sideways and that joints 2, 4 and 6 are not allowed to rotate. This implies that there must somehow be (a) an external horizontal force preventing sway, and (b) an external clamping moment at each of the joints preventing rotation. These external actions do not exist in the real structure, of course, but we can imagine that by some means they have been applied temporarily to prevent displacement. At this stage of the calculation the moments acting on the ends of the members are the fixed-end moments, which are

$$M^F_{24} = -\frac{10 \times 24^2}{12} = -480 \qquad M^F_{42} = +480$$

$$M^F_{43} = -\frac{80 \times 15}{8} = -150 \qquad M^F_{34} = +150$$

$$M^F_{46} = -\frac{200 \times 24}{8} = -600 \qquad M^F_{64} = +600$$

Now consider joint 2. The total moment acting on the joint will be the sum of the fixed-end moment $-M^F_{24} = +480$ and the external clamping moment which is preventing rotation.

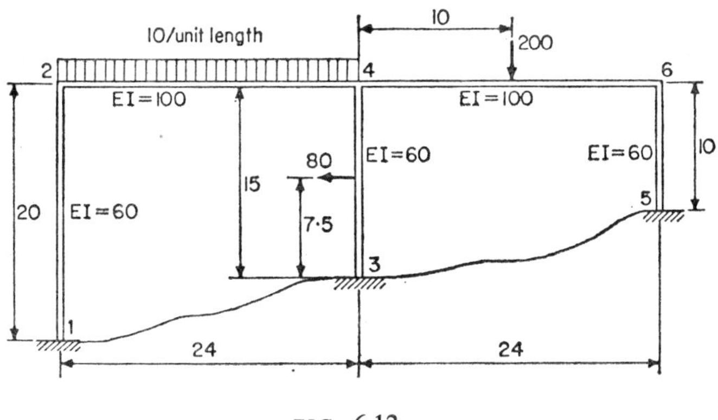

FIG. 6.12

The clamping moment is now removed and the joint is allowed to rotate freely until the moments due to the bending of the two beams exactly cancel out the out of balance moment $-M^F_{24}$. If the rotation of the joint is θ_2, then the moments acting on the joint due to this will be

$$\begin{aligned} -M_{21} &= -S_{21}\theta_2 \\ -M_{24} &= -S_{24}\theta_2 \end{aligned} \tag{6.2}$$

and the condition for equilibrium will be

$$M^F_{24} + M_{21} + M_{24} = 0$$

From (6.2) this may be written as

$$M^F_{24} + (S_{21} + S_{24})\theta_2 = 0 \tag{6.3}$$

The angle θ_2 may now be eliminated from equations (6.2) and (6.3) and the beam-end moments M_{21} and M_{24} written as

$$M_{21} = -\frac{S_{21}}{S_{21} + S_{24}} M^F_{24}$$

$$M_{24} = -\frac{S_{24}}{S_{21} + S_{24}} M^F_{24}$$

Thus when the clamping moment is relaxed and the joint is allowed to rotate freely the moment applied to the joint is distributed to the members connected to the joint in the ratio of their stiffness factors. The equations may also be written as

$$M_{21} = -D_{21}M^F_{24}$$
$$M_{24} = -D_{24}M^F_{24}$$

where in general D_{ij} is called a *distribution factor* and is defined by

$$D_{ij} = \frac{S_{ij}}{\sum\limits_{\text{joint } i} S_{ik}}$$

and where $\sum\limits_{\text{joint } i} S_{ik}$ is the sum of the stiffness factors at joint i. In the case of joint 2,

$$S_{21} = \frac{4 \times 60}{20} = 12{\cdot}00$$

$$S_{24} = \frac{4 \times 100}{24} = 16{\cdot}67$$

so that

$$D_{21} = \frac{12{\cdot}00}{28{\cdot}67} = 0{\cdot}419$$

$$D_{24} = \frac{16{\cdot}67}{28{\cdot}67} = 0{\cdot}581$$

Note that the sum of the distribution factors at a node must always be unity. The moments due to bending become

$$M_{21} = -0{\cdot}419 \times (-480) = +201$$
$$M_{24} = -0{\cdot}581 \times (-480) = +279$$

and equilibrium at joint 2 is achieved, for the sum of all the moments acting on the joint is

$$-480 + 201 + 279 = 0$$

At this stage joint 2 is clamped again to prevent further rotation while the process of relaxation and distribution of moments is applied to the other joints in turn. First, the stiffness and distribution factors are calculated:

$$S_{42} = \frac{4 \times 100}{24} = 16{\cdot}67 \qquad D_{42} = \frac{16{\cdot}67}{49{\cdot}34} = 0{\cdot}338$$

$$S_{43} = \frac{4 \times 60}{15} = 16{\cdot}00 \qquad D_{43} = \frac{16{\cdot}00}{49{\cdot}34} = 0{\cdot}324$$

$$S_{46} = \frac{4 \times 100}{24} = 16{\cdot}67 \qquad D_{46} = \frac{16{\cdot}67}{49{\cdot}34} = 0{\cdot}338$$

$$\overline{49{\cdot}34} \qquad \overline{1{\cdot}000}$$

$$S_{64} = \frac{4 \times 100}{24} = 16{\cdot}67 \qquad D_{64} = \frac{16{\cdot}67}{40{\cdot}67} = 0{\cdot}410$$

$$S_{65} = \frac{4 \times 60}{10} = 24{\cdot}00 \qquad D_{65} = \frac{24{\cdot}00}{40{\cdot}67} = 0{\cdot}590$$

$$\overline{40{\cdot}67} \qquad \overline{1{\cdot}000}$$

The distribution of moments at joints 4 and 6 and the subsequent calculations are best carried out using the tabular scheme of Table 6.1. It is an essential part of the technique of moment distribution that a consistent tabular scheme is adopted, and it is important to make the entries in a neat and systematic manner. The actual form of the table depends on preference and convenience. If the calculation is being carried out on a drawing board, for instance, it would generally be better to sketch the outline of the structure and enter the various moments on the drawing at the ends of the relevant members. In Table 6.1 it can be seen that the calculations are roughly laid out in the same shape as the structure with the columns of figures containing moments at joints 1, 3 and 5 occurring beneath those for joints 2, 4 and 6.

The first row in the table specifies the members at each joint, and the next two give the distribution factors and carry-over factors for ease of reference. At joints 1, 3 and 5 the distribution factors are zero as the columns are built in to the foundation, while for the same reason the carry-over factors at these joints have been omitted. The fourth row begins the moment distribution process proper and contains the fixed-end moments: this row is labelled 'line 1' on the left-hand side. The distribution of the fixed-end moment of

TABLE 6.1 No-Sway Moments, M_{ij}^{0}

Line	Joint 2		Joint 4			Joint 6		
	2–1	2–4	4–2	4–3	4–6	6–4	6–5	Member
	0·419	0·581	0·338	0·324	0·338	0·410	0·590	Distribution factor
	½ ↓	½→	←½	½ ↓	½→	←½	½ ↓	Carry-over factor
1	0	−480	+480	−150	−600	+600	0	Fixed-end moments
2	+201	+279	+91	+88	+91	−246	−354	Distributed moments
3	0	+45	+139	0	−123	+45	0	Carried-over moments
4	−19	−26	−5	−5	−6	−18	−27	Distributed moments
5	0	−2	−13	0	−9	−3	0	Carried-over moments
6	+1	+1	+7	+7	+8	+1	+2	Distributed moments
7	0	+3	0	0	0	+4	0	Carried-over moments
8	−1	−2	0	0	0	−2	−2	Distributed moments
9	+182	−182	+699	−60	−639	+381	−381	Total no-sway moments

Line	Joint 1	Joint 3	Joint 5	
	1–2	3–4	5–6	Member
	0	0	0	Distribution factor
1	0	+150	0	Fixed-end moments
2	0	0	0	Distributed moments
3	+100	+44	−177	Carried-over moments
4	0	0	0	Distributed moments
5	−9	−2	−13	Carried-over moments
6	0	0	0	Distributed moments
9	+91	+192	−190	Total no-sway moments

-480 at joint 2 has already been calculated, and the distributed moments of $+201$ and $+279$ are entered into line 2 in the appropriate columns. Three fixed-end moments occur at joint 4. Their total value is -270, and this moment is distributed into members 4–2, 4–3, and 4–6 and entered into line 2. Similarly at joint 6, the fixed-end moment of 600 is distributed according to the distribution factors D_{64} and D_{65} and entered into the table. It should be noted that at this stage, the total moments acting at any joint in lines 1 and 2 add up to zero: each joint is in equilibrium, and a horizontal line is drawn in the table to indicate this.

Unfortunately, however, when joint 2 is allowed to rotate and moments of $+201$ and $+279$ are distributed into members 2–1 and 2–4, moments are also induced at the far ends of the members by the rotation. The moment induced at 4 is equal to the moment at 3 multiplied by the carry-over factor of $\frac{1}{2}$, and is equal to

$$279 \times \tfrac{1}{2} = 139$$

The figure is rounded to the nearest whole number as the introduction of additional significant figures would not improve the accuracy at this stage. The carried-over moment is now entered in line 3 under column 4–2. The arrows in the table indicate the directions of the carry-over. The moment of 201 at the end of 2–1 will similarly carry-over a moment of $201 \times \frac{1}{2} = 100$ to 1–2, where again it appears in line 3. The remaining moments in line 2 are also carried over, and line 3 is completed.

At this stage the joints are again not in a state of equilibrium, though the moments in line 3 are considerably less than the original ones in line 1. Out of balance moments exist which must be removed by distributing them into the members as before (line 4) by relaxing each joint in turn. A horizontal line is again drawn to indicate that the joints are in equilibrium, and the moments in line 4 are carried-over into line 5. This process of distributing and carrying-over is continued until the out-of-balance moments are sufficiently small, and then the moments in each column are added, as shown in line 9. These moments are still not the correct joint moments of the structure, for the calculation has assumed the frame nodes are fixed in space and are not free to move laterally. This restriction can lead to considerable errors, so a correction to allow for the sidesway of the structure must now be made.

6.5 Sway Correction

The moment distribution procedure for handling sway problems is very similar to that used with slope-deflection except that once again the displacement unknown is not found explicitly but rather, the whole solution is carried out in terms of moments.

To begin the calculation of the sidesway effects, all joints are clamped against rotation, the applied load is removed and the structure is given a lateral displacement Δ as in fig. 6.13. This displacement is the correct, final sway displacement for the structure although its value is not, of course, known explicitly at this stage. Bending moments are caused in all the columns: let the moments at the top and bottom of member 1–2 be

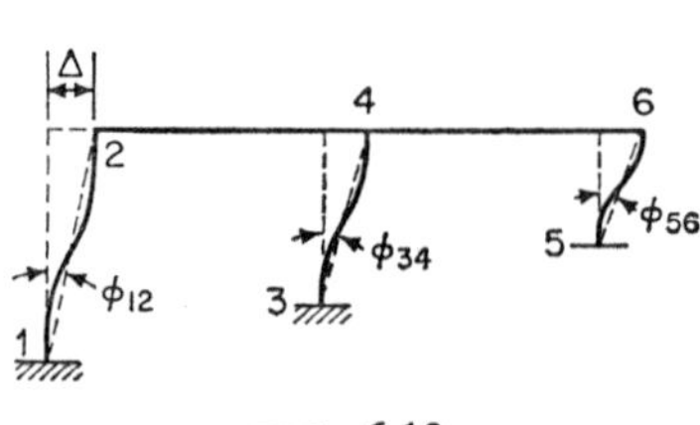

FIG. 6.13

$$M_{21} = M_{12} = 1000f$$

where f is some factor to be determined later by considering the equilibrium of the structure, and where the choice of the number 1000 is quite arbitrary.

The relevant part of the slope-deflection equation for member 1–2 is

$$M_{12} = -(S_{12} + C_{21}S_{21})\varphi_{12}$$

$$= -S_{12}(1 + C_{12})\varphi_{12} = -\frac{6EI}{L_{21}}\varphi_{12}$$

Similarly

$$M_{34} = -\frac{6EI}{L_{34}}\varphi_{34}$$

Now by geometry,

$$\varphi_{34} = \frac{20}{15}\varphi_{12}$$

Hence in terms of the factor f

$$M_{34} = +\frac{6EI}{15} \times \frac{20}{15} \times \frac{20}{6EI} \times 1000f = \frac{20^2}{15^2} \times 1000f = 1780f$$

and similarly

$$M_{56} = M_{65} = \frac{20^2}{10^2} \times 1000f = 4000f$$

These moments are called the *initial-sway moments*. Note that for a one-degree-of-freedom system such as the one under consideration, once one of the initial-sway moments has been assigned a value (in this case M_{12}), all the other initial-sway moments are uniquely determined by the geometry of the structure.

A second moment distribution is now carried out in Table 6.2. The distribution and carry-over factors are the same, but this time the initial out-of-balance moments at the joints are obtained from the initial-sway moments rather than from fixed-end moments due to applied loads.

TABLE 6.2 Sway Moments, M^1_{ij}

Joint 2		Joint 4			Joint 6	
2–1	2–4	4–2	4–3	4–6	6–4	6–5
0·419	0·581	0·338	0·324	0·338	0·410	0·590
½↓	½→	←½	½↓	½→	←½	½↓
+1000 −419	0 −581	0 −602	+1780 −576	0 −602	0 −1640	+4000 −2360
0 +126	−301 +175	−290 +375	0 +360	−820 +375	−301 +124	0 +177
0 −79	+188 −109	+88 −51	0 −48	+62 −51	+188 −77	0 −111
0 10	−25 +15	−54 +31	0 +30	−38 +31	−25 +10	0 +15
+638	−638	−503	+1546	−1043	−1721	+1721
Joint 1			Joint 3			Joint 5
1–2			3–4			5–6
+1000 0			+1780 0			+4000 0
−210 0			−288 0			−1180 0
+63 0			+180 0			+88 0
−40			−24			−55
+813			+1648			+2853

The factor f is now found from the conditions of equilibrium of the structure, and for this purpose a moment-sway equation is written. The formation of moment-sway equations was discussed in Section 6.3, where it was shown that the simplest way of setting up such an equation was by the use of the principle of virtual work. In the present case the kinematic relations between the chord rotations have already been calculated, but to repeat them and write them in the form of small variations,

$$\delta\varphi_{34} = \tfrac{20}{15}\delta\varphi_{21} = 1{\cdot}333\delta\varphi_{21}$$
$$\delta\varphi_{56} = \tfrac{20}{10}\delta\varphi_{21} = 2\delta\varphi_{21}$$

The virtual displacements applied to the structure are shown in fig. 6.14. The moment-sway equation becomes

$$(M_{12} + M_{21})\,\delta\varphi_{12} + (M_{34} + M_{43})\,\delta\varphi_{34} + (M_{56} + M_{65})\,\delta\varphi_{56} - 80 \times \tfrac{15}{2}\delta\varphi_{34} = 0$$

where $M_{12}, M_{21} \ldots$ are the actual moments on the structure comprising both the no-sway moments $M^0_{12}, M^0_{21} \ldots$ from Table 6.1 and the sway moments $fM^1_{12}, fM^1_{21} \ldots$ calculated in Table 6.2, so that

$$M_{12} = M^0_{12} + fM^1_{12}$$

and so on. It should be noticed that the lateral force of 80 on member 3–4 does work as the structure is displaced and so must be included in the virtual

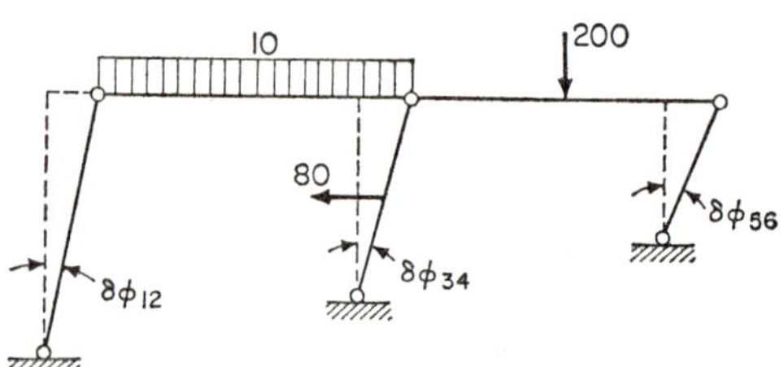

FIG. 6.14 Virtual displacements

work equation. Rewriting the moment-sway equation in terms of components of moments and eliminating the virtual displacements gives

$$M^0_{12} + M^0_{21} + 1{\cdot}333(M^0_{34} + M^0_{43}) + 2M^0_{56} + 2M^0_{65} + f[M^1_{12} + M^1_{21} + 1{\cdot}333(M^1_{34} + M^1_{43}) + 2M^1_{56} + 2M^1_{65}] - 800 = 0$$

Substituting the values found in Tables 6.1 and 6.2 into this equation leads to

$$182 + 91 + 1{\cdot}333(-60 + 192) - 762 - 380 + f[638 + 813 + 1{\cdot}333(1546 + 1648) + 3442 + 5706] - 800 = 0$$

The value of f is then found to be

$$f = +0{\cdot}1006$$

The final moments may now be found by multiplying the sway correction moments by f and adding them to the no-sway moments, as in the following table:

Moment	M_{12}	M_{21}	M_{42}	M_{43}	M_{46}	M_{34}	M_{65}	M_{56}
Uncorrected sway moment	+813	+638	−503	+1546	−1043	+1648	+1721	+2853
Corrected sway moment	+82	+64	−51	+155	−105	+166	+173	+287
No-sway moment	+91	+182	+699	−60	−639	+192	−381	−190
Total moment	+173	+246	+648	+95	−744	+358	−208	+97

The bending moment diagram for the structure is shown in fig. 6.15.

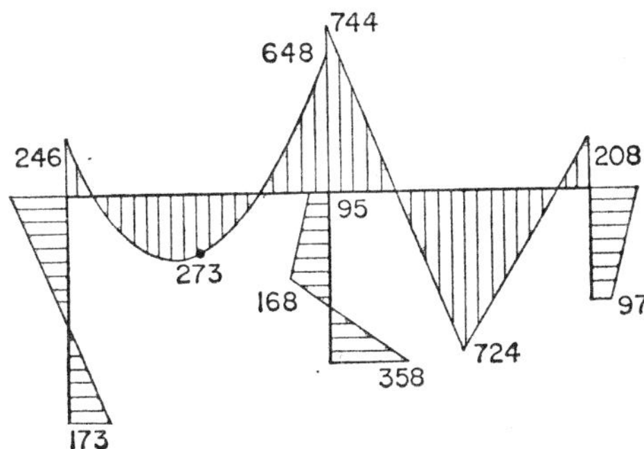

FIG. 6.15

6.6 Symmetry

Symmetrical structures were discussed in Section 5.5, in which it was shown that any load distribution on a symmetric structure could be broken down into a symmetric and an antisymmetric loading, and that the calculations for a force-method solution could be considerably simplified by making use of the symmetry of a structure. Displacement solutions may also be simplified in this way: only half the structure need be considered, and it often happens for simple structures that a moment distribution solution can be completed with only one distribution. To illustrate the treatment of a symmetric structure the example solved in Section 5.5 is analysed again as follows.

First, the loading on the structure (fig. 6.16(a)) is split into symmetric and antisymmetric components (b), and half the structure is taken with appropriate boundary conditions at the centreline (c). For the symmetric half-structure point 3 can move vertically but the structure cannot sway sideways, while the support at 3 for the antisymmetric half-structure is free both to rotate and to translate laterally. The fixed-end moments for the two cases may

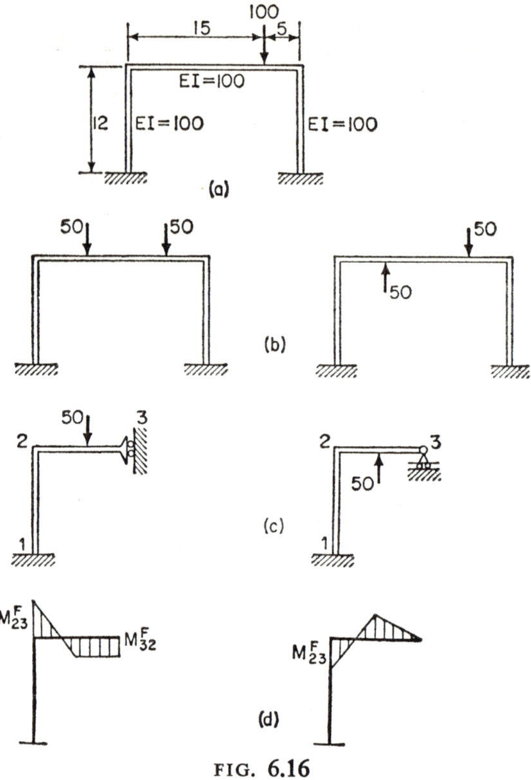

FIG. 6.16

now be found by considering the bending moment diagrams of fig. 6.16(d). For the symmetric case the total area of the bending moment diagram is zero as there is no rotation either at 2 or at 3. From this and the geometry of the diagram, the fixed-end moments can be found to be

$$M^F_{23} = -187{\cdot}5; \qquad M^F_{32} = -62{\cdot}5$$

The result for the antisymmetric case can best be found by a modification of a known result. If joint 3 were fixed, as well as 2, then the moments at 2 and 3 would be

$$M^F_{32} = -M^F_{23} = -\frac{50 \times 10}{8} = -62{\cdot}5$$

If joint 3 were now allowed to relax and to rotate freely, this would superimpose a moment of $+62{\cdot}5$ at 3 to make the total moment zero, which would carry over a moment of $+\frac{1}{2} \times 62{\cdot}5$ to joint 2. The antisymmetric fixed-end moment at 2 would then be

$$M_{23}^{F} = +62{\cdot}5 + \tfrac{1}{2} \times 62{\cdot}5 = +93{\cdot}75$$

The stiffness and distribution factors for the symmetric case are, remembering that the roller end condition at 3 calls for a stiffness factor of EI/L,

$$S_{21} = \frac{4 \times 100}{12} = 33{\cdot}3 \qquad D_{21} = 0{\cdot}769$$

$$S_{23} = \frac{100}{10} = 10 \qquad D_{23} = 0{\cdot}231$$

The carry-over factors are

$$C_{21} = \tfrac{1}{2}; \qquad C_{23} = -1$$

The moment distribution for the symmetric case is carried out in Table 6.3.

TABLE 6.3 Symmetric Distribution

2–1	2–3	3–2
0·769	0·231	0
$\frac{1}{2}\downarrow$	$-1\rightarrow$	0
0 +144·1	−187·5 +43·4	−62·5 0
0	0	−43·4
+144·1	−144·1	−105·9

1–2
0
0 0
+72·0
+72·0

It should be noted that once the first distribution has been made, no further distribution is possible.

For the antisymmetric case, the stiffness and distribution factors are

$$S_{21} = \frac{4 \times 100}{12} = 33{\cdot}3 \qquad D_{21} = \frac{33{\cdot}3}{63{\cdot}3} = 0{\cdot}526$$

$$S_{23} = \frac{3 \times 100}{10} = 30 \qquad D_{32} = 0{\cdot}474$$

The structure is evidently going to move sideways, so that both a no-sway and a sway distribution must be carried out: these distributions are set out in Table 6.4. Because the only rotational degree of freedom occurs at joint 2,

TABLE 6.4 Antisymmetric Distribution

(a) No-Sway Distribution

2–1	2–3
0·526	0·474
½↓	0
0 −49·2	+93·7 −44·5
0	0
−49·2	+49·2

1–2
0
0 0
−24·6
−24·6

(b) Sway Distribution

2–1	2–3	
0·526	0·474	
½↓	0	
+100 −52·6	0 −47·4	
0	0	
+47·4	−47·4	Arbitrary moments
+28·9 −49·2		Corrected sway moments No-sway moments
−20·3		Total antisymmetric moments

1–2
0
+100 0
−26·3
+73·7
+44·9 −24·6
+20·3

once again a single distribution completes the equilibrium of the entire frame and no iteration is necessary. The initial-sway moments for the sway distribution only occur in member 2–1, so that an arbitrary choice of a moment of 100 may be made at the top and bottom of the column. The moment-sway equation for the antisymmetric case is, in terms of a virtual rotation $\delta\varphi$ of the column,

$$M_{21}\,\delta\varphi + M_{12}\,\delta\varphi = 0$$

or simply

$$M_{21} + M_{12} = 0$$

Now, the total moment at any point is made up of the no-sway moment and the sway moment, or, for example,

$$M_{12} = M_{12}^{0} + M_{12}^{1}f$$

the moment-sway equation therefore becomes

$$-49{\cdot}2 - 24{\cdot}6 + (47{\cdot}4 + 73{\cdot}7)f = 0$$

so that

$$f = +0{\cdot}609$$

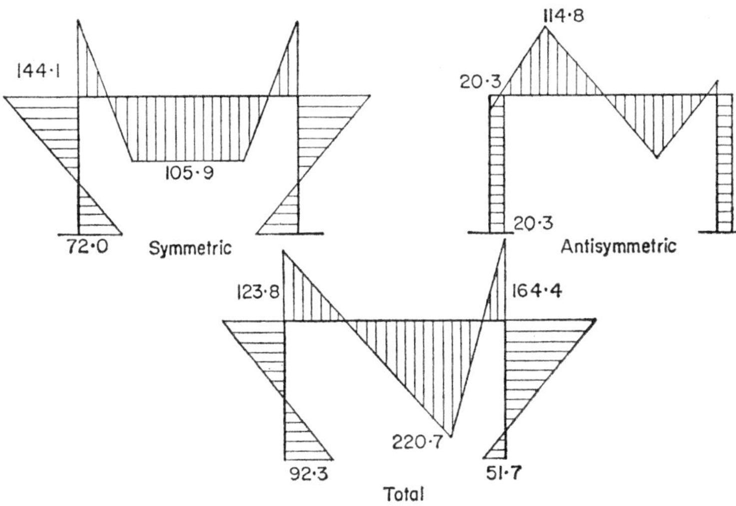

FIG. 6.17 Bending moment diagrams

The arbitrary sway moments are multiplied by this factor in Table 6.4(b) and added to the no-sway moments to get the total antisymmetric moments. The symmetric and antisymmetric bending moment diagrams are drawn in fig. 6.17 and combined to give the total bending moment diagram for the structure.

6.7 Flexible Supports

Although it is usually justifiable to neglect support flexibility in structural analysis, there are some instances in which the effect of support deflections is too important to be omitted. The structure shown in fig. 6.18 is such a case, for the deflections of the transverse beam 10–11 on which the centre of the frame rests affect the behaviour of the structure very considerably. This is a fairly typical flexible-support problem: the extreme instance of this type of structure is the grillage, in which a number of beams are connected at right angles to one another.

The problem in fig. 6.18 is symmetric so that only half the structure need be considered. The loading, too, is symmetric so that there will be no sidesway: node 5 will, however, move vertically, and this degree of freedom will be treated in the same way as sidesway by first fixing it against displacement for a no-sway distribution, and then carrying out a sway distribution in terms of arbitrary moments.

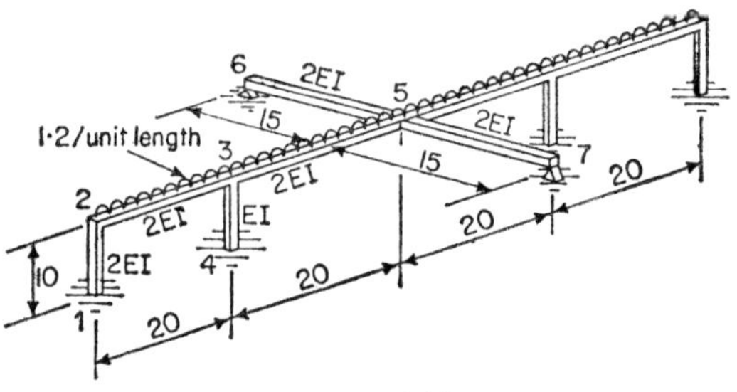

FIG. 6.18

With all joints fixed against rotation and joint 5 fixed against translation, the ends of the beams will all be subjected to the same fixed-end moment, represented typically by

$$M_{23}^{F} = -\frac{20^2 \times 1{\cdot}2}{12} = -40$$

The absolute value of the section stiffness *EI* does not come into the calculations, so taking an arbitrary value for *EI* of 10, the stiffness factors at joints 2 and 3 become

$$S_{21} = \frac{4 \times 20}{10} = 8 \qquad S_{32} = \frac{4 \times 20}{20} = 4$$

$$S_{23} = \frac{4 \times 20}{20} = 4 \qquad S_{34} = \frac{4 \times 10}{10} = 4$$

$$S_{35} = \frac{4 \times 20}{20} = 4$$

so that the distribution factors are therefore

$$D_{21} = 0{\cdot}667 \qquad D_{32} = 0{\cdot}333$$
$$D_{23} = 0{\cdot}333 \qquad D_{34} = 0{\cdot}334$$
$$D_{35} = 0{\cdot}333$$

The no-sway distribution is carried out as follows, paying no attention to the moments at the feet of the frame which are irrelevant at this stage:

TABLE 6.5

2–1	2–3	3–2	3–4	3–5	5–3
0·667	0·333	0·333	0·334	0·333	0
$\frac{1}{2}\downarrow$	$\frac{1}{2}\rightarrow$	$\leftarrow\frac{1}{2}$	$\frac{1}{2}\downarrow$	$\frac{1}{2}\rightarrow$	—
0 +26·7	−40·0 +13·3	+40·0 0	0 0	−40·0 0	+40·0 0
0 0	0 0	+6·6 −2·2	0 −2·2	0 −2·2	0 0
0 +0·7	−1·1 +0·4	0 0	0 0	0 0	−1·1 0
+27·4	−27·4	+44·4	−2·2	−42·2	+38·9

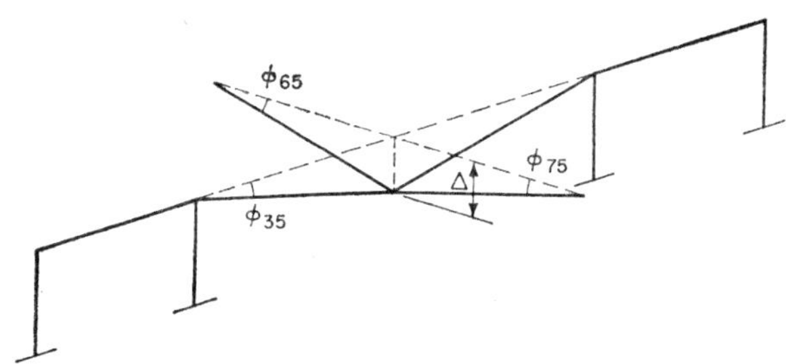

FIG. 6.19 Sway mechanism

The applied load is now removed and node 5 is displaced downwards by some amount Δ with joints 2 and 3 fixed against rotation. The equivalent sway mechanism is shown in fig. 6.19, and from the geometry of this diagram it can be seen that the chord rotation relationship is

$$15\varphi_{65} = 20\varphi_{35}$$
$$\varphi_{65} = 1{\cdot}333\varphi_{35}$$

If the initial-sway moments induced in 3–5 are taken to be

$$M_{35} = M_{53} = -1000f$$

then as

$$M_{35} = -S_{35}(1 + C_{35})\varphi_{35}$$
$$= -6\varphi_{35}$$

and

$$M_{56} = -S_{56}(1 + C_{56})\varphi_{56}$$
$$= -\frac{3 \times 20}{15}(1 + 0)\varphi_{56} = -4\varphi_{56}$$

the initial-sway moment in the transverse beam becomes

$$M_{56} = -\frac{4}{6} \times 1{\cdot}333 \times 1000f = -889f$$

The sway distribution is now carried out in Table 6.6.

TABLE 6.6

2–1	2–3	3–2	3–4	3–5	5–3
0·667	0·333	0·333	0·334	0·333	0
$\frac{1}{2}\downarrow$	$\frac{1}{2}\rightarrow$	$\leftarrow\frac{1}{2}$	$\frac{1}{2}\downarrow$	$\frac{1}{2}\rightarrow$	—
0	0	0 +333	0 +334	−1000 +333	−1000
 −111	+167 −56				+167
		−28 +10	+9	+9	
 −3	+5 −2				+4
−114	+114	+315	+343	−658	−829

The final moments produced by this distribution are, of course, the uncorrected sway moments and must be multiplied by a factor f whose value can be found by solving a moment-sway equilibrium equation.

In setting up the moment-sway equation based on the virtual displacements of the mechanism of fig. 6.19, it is important to sum the virtual work done by the whole structure. The work done on 3–5 must therefore be doubled, and the virtual work equation is

$$2 \times (M_{35} + M_{53})\,\delta\varphi_{35} + M_{56}\,\delta\varphi_{56} + M_{57}\,\delta\varphi_{57} + \frac{2 \times 1{\cdot}2 \times 20^2}{2}\,\delta\varphi_{35} = 0$$

which becomes, on using the chord rotation relations and remembering that the work done on each half of the transverse beam must be the same,

$$2(M_{35} + M_{53}) + 2 \times 1{\cdot}333M_{56} + 480 = 0$$

As in the previous example, each moment is composed of a no-sway component (although the no-sway value of M_{56} is zero), so that on substituting the values calculated in the distribution tables the equation becomes

$$2(-42{\cdot}2 + 38{\cdot}9) + 480 + [2(-658 - 829) - 2 \times 1{\cdot}333 \times 889]f = 0$$

Hence

$$f = 0{\cdot}0886$$

The final moments may now be calculated in the following table.

Moment	M_{12}	M_{21}	M_{32}	M_{34}	M_{43}	M_{35}	M_{53}	M_{56}
Arbitrary sway moment		−114	+315	+343		−658	−829	−889
Corrected sway moment		−10·1	+27·9	+30·4		−58·3	−73·5	−78·8
No-sway moment		+27·4	+44·4	−2·2		−42·2	+38·9	0
Total moment	+8·6	+17·3	+72·3	+28·2	+14·1	−100·5	−34·6	−78·8

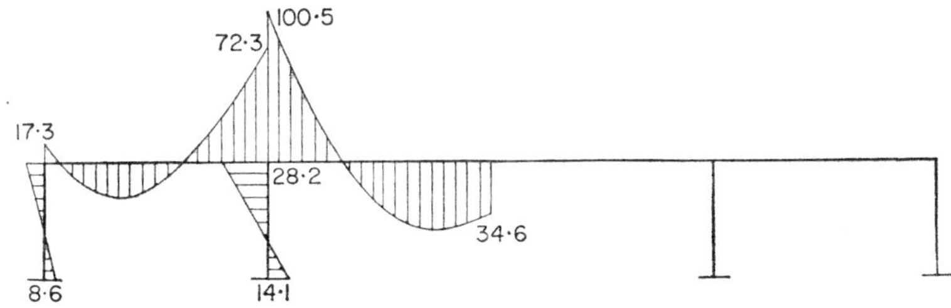

FIG. 6.20

The moments at the bottoms of the columns are merely found by multiplying M_{21} and M_{43} by the carry-over factor. The bending moment diagram is plotted in fig. 6.20.

6.8 Inclined Frames

All the structures so far considered in this chapter have been rectangular frames whose members have been either vertical or horizontal. If a frame has inclined members, it can be solved in exactly the same way, using the same principles. The only difference is that the kinematic relations between the chord rotations become somewhat more complex, making the initial-sway moments more awkward to set up and the moment-sway equations rather lengthier.

A typical bent with inclined members is shown in fig. 6.21. The chord rotation relationships can be found in one of two ways. The first is to calculate the nodal displacements in terms of chord rotations and then to eliminate the displacements from the resulting kinematic equations, a procedure which is best carried out by expressing the node positions in terms of their horizontal

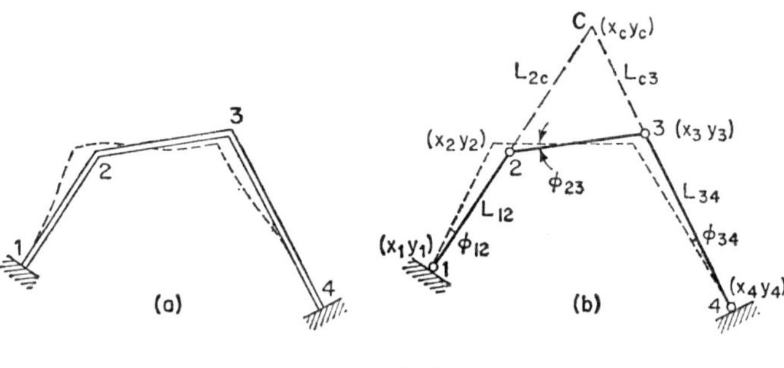

FIG. 6.21

and vertical coordinates. The equivalent sway mechanism for the frame is shown in fig. 6.21(b). If member 1–2 rotates through a small angle φ_{12}, then the horizontal and vertical displacements u_2 and v_2 of joint 2 are

$$u_2 = \varphi_{12}(y_2 - y_1)$$
$$v_2 = -\varphi_{12}(x_2 - x_1)$$

The rotation φ_{23} of member 2–3 (which is, as usual, assumed to be clockwise) gives relative displacements between nodes 2 and 3 of

$$u_3 - u_2 = \varphi_{23}(y_3 - y_2)$$
$$v_3 - v_2 = -\varphi_{23}(x_3 - x_2)$$

and the rotation of 3–4 gives

$$u_3 = \varphi_{34}(y_3 - y_4)$$
$$v_3 = -\varphi_{34}(x_3 - x_4)$$

Eliminating u_2, u_3, v_2 and v_3 from these equations, we get

$$\varphi_{34}(y_3 - y_4) - \varphi_{12}(y_2 - y_1) = \varphi_{23}(y_3 - y_2)$$
$$\varphi_{34}(x_3 - x_4) - \varphi_{12}(x_2 - x_1) = \varphi_{23}(x_3 - x_2)$$

from which the required chord rotation relationships can be found. It is, however, inadvisable to try to remember these formulae and use them as they stand. Rather, their derivation should be regarded as a general illustration of the method of finding chord rotation relationships, and each individual case should be treated on its own merits.

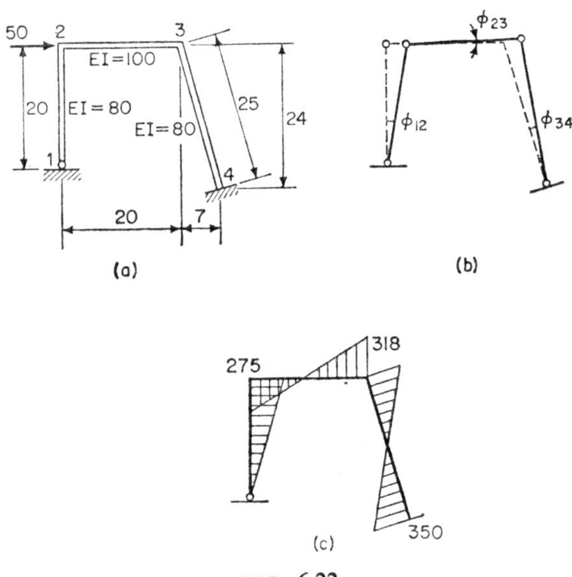

FIG. 6.22

To illustrate this, consider the particular case shown in fig. 6.22. The horizontal displacements of 2 and 3 are the same so that we can immediately write down that

$$u_2 = 20\varphi_{12} = u_3 = 24\varphi_{34}$$

The vertical displacement of 3 is

$$v_3 = -20\varphi_{23} = +7\varphi_{34}$$

Hence

$$\varphi_{12} = 1{\cdot}2\varphi_{34}$$
$$\varphi_{23} = -0{\cdot}35\varphi_{34}$$

The second way of finding chord rotation relationships is to find the instantaneous centre of rotation of member 2–3, shown as point C in fig. 6.21(b). For a small displacement of the mechanism point 2 moves perpendicular to 1–2, and as the radius from C to 2 must itself be perpendicular to the direction of movement of 2, C must lie on the line of 1–2. It must similarly lie on the line of 3–4, and its position is therefore given by the intersection of the two lines. Triangle 2C3 should be thought of as a rigid whole rotating about C. The relation between the chord rotations of 1–2 and 2–3 can then be written in one of three ways, namely

$$\varphi_{12}(x_2 - x_1) = -\varphi_{23}(x_C - x_2)$$

or

$$\varphi_{12}(y_2 - y_1) = -\varphi_{23}(y_C - y_2)$$

or

$$\varphi_{12}L_{12} = -\varphi_{23}L_{2C}$$

In a particular problem, the most convenient of the three is used. Similarly

$$\varphi_{34}(y_3 - y_4) = -\varphi_{23}(y_C - y_3)$$
$$\varphi_{34}(x_3 - x_4) = -\varphi_{23}(x_C - x_3)$$
$$\varphi_{34}L_{34} = -\varphi_{23}L_{C3}$$

Again using the frame of fig. 6.22 to illustrate the method, suppose the instantaneous centre of rotation of 2–3 is denoted by C (not shown in the figure). By similar triangles, the length of 2–C is

$$L_{2C} = \frac{24 \times 20}{7} = 68{\cdot}6$$

Hence

$$\varphi_{12} = -\frac{68{\cdot}6}{20}\varphi_{23} = -3{\cdot}43\varphi_{23}$$

and

$$\varphi_{34} = -\frac{68{\cdot}6}{24}\varphi_{23} = -2{\cdot}86\varphi_{23}$$

The choice of method for finding the chord rotation relationships depends entirely on how easy it is to find the instantaneous centre of rotation. If its position can be determined without much difficulty, then it is the better method to use; otherwise the more straightforward node-displacement method would be preferable.

The analysis of the structure of fig. 6.22 will now be completed using the slope-deflection method. There are no fixed-end moments to be calculated as the only load on the structure is applied at a node point. The stiffness factors are

$$S'_{21} = \frac{3 \times 80}{20} = 12$$

where the factor has been modified to allow for the hinge at 1,

$$S_{23} = S_{32} = \frac{4 \times 100}{20} = 20$$

and

$$S_{34} = S_{43} = \frac{4 \times 80}{\sqrt{(24^2 + 7^2)}} = 12{\cdot}8$$

The slope-deflection equations are therefore, from equation (6.1),

$$\begin{aligned}
M_{12} &= 0 \\
M_{21} &= 12\theta_2 - 12\varphi_{12} \\
M_{23} &= 20\theta_2 + 10\theta_3 - (20 + 10)\varphi_{23} \\
M_{32} &= 10\theta_2 + 20\theta_3 - (20 + 10)\phi_{23} \\
M_{34} &= 12{\cdot}8\theta_3 - (12{\cdot}8 + 6{\cdot}4)\varphi_{34} \\
M_{43} &= 6{\cdot}4\theta_3 - (12{\cdot}8 + 6{\cdot}4)\varphi_{34}
\end{aligned}$$

in which, of course, we can write φ_{12} and φ_{34} in terms of φ_{23}. The two joint-equilibrium equations are

$$\begin{aligned}
M_{21} + M_{23} &= 0 \\
M_{32} + M_{34} &= 0
\end{aligned}$$

While using the sway mechanism of fig. 6.22(b) as the basis of a virtual work equation, the moment-sway equilibrium equation becomes

$$M_{21}\,\delta\varphi_{21} + (M_{23} + M_{32})\,\delta\varphi_{23} + (M_{34} + M_{43})\,\delta\varphi_{34} + 50 \times 20\delta\varphi_{21} = 0 \qquad (6.4)$$

or, using the known chord rotation relations,

$$-3{\cdot}43M_{21} + M_{23} + M_{32} - 2{\cdot}86(M_{34} + M_{43}) - 3{\cdot}43 \times 1000 = 0$$

In terms of the unknown displacements the three equilibrium equations become

$$\begin{aligned}
&(12 + 20)\theta_2 + 10\theta_3 + (12 \times 3{\cdot}43 - 30)\,\varphi_{23} = 0 \\
&10\theta_2 + (20 + 12{\cdot}8)\theta_3 - (30 - 2{\cdot}86 \times 19{\cdot}2)\,\varphi_{23} = 0 \\
&(-3{\cdot}43 \times 12 + 20 + 10)\theta_2 + (10 + 20 - 2{\cdot}86 \times 19{\cdot}2)\theta_3 \\
&\quad + (-3{\cdot}43 \times 3{\cdot}43 \times 12 - 30 - 2{\cdot}86 \times 2{\cdot}86 \times 19{\cdot}2 \times 2)\,\varphi_{23} \\
&\quad - 3430 = 0
\end{aligned}$$

In matrix form these equations become

$$\begin{bmatrix} 30 & 10 & 11{\cdot}16 \\ 10 & 32{\cdot}8 & 24{\cdot}9 \\ 11{\cdot}16 & 24{\cdot}9 & 515 \end{bmatrix} \begin{Bmatrix} \theta_2 \\ \theta_3 \\ \varphi_{23} \end{Bmatrix} = \begin{Bmatrix} 0 \\ 0 \\ 3430 \end{Bmatrix}$$

The solution is

$$\begin{aligned} \varphi_{23} &= -6{\cdot}95 \\ \theta_3 &= +5{\cdot}0 \\ \theta_2 &= +0{\cdot}918 \end{aligned}$$

Hence, back-substituting into the slope-deflection equations,

$$\begin{aligned} M_{21} &= -275 \\ M_{34} &= -318 \\ M_{43} &= -350 \end{aligned}$$

These moments are plotted in fig. 6.22(c).

The solution of the problem by moment distribution follows similar lines. There are no fixed-end moments so that a no-sway distribution is not needed. For the sway distribution, the initial-sway moments must be set up with care. If the choice of an initial arbitrary moment is

$$M_{23} = M_{32} = 1000f$$

then the other initial-sway moments may be found by using the general slope-deflection equation with the joint rotations set at zero: the joints cannot be allowed to rotate before distribution commences. Then

$$\begin{aligned} M_{23} &= 1000f = -(S_{23} + C_{32}S_{32})\varphi_{23} = -30\varphi_{23} \\ M_{21} &= -S'_{21}\varphi_{21} = 12 \times 3{\cdot}43\varphi_{23} \\ &= -3{\cdot}43 \times 12 \times \frac{1000}{30}f = -1372f \\ M_{34} &= -2{\cdot}86 \times (12{\cdot}8 + 6{\cdot}4) \times \frac{1000}{30}f = -1830f \end{aligned}$$

The distribution of the initial-sway moments is carried out in Table 6.7, in which the distributed arbitrary moments are summed in the penultimate line. The moment-sway equation is of course the same as equation (6.4) used in the slope-deflection solution. Substituting the uncorrected moments in this gives

$$[-3{\cdot}43 \times (-1302) + 1302 + 1527 - 2{\cdot}86(-1527 - 1679)]f - 3430 = 0$$

from which

$$f = +0{\cdot}208$$

This factor is applied to the uncorrected moments to produce the corrected moments in the last line of Table 6.7. These results agree with those calculated by the slope-deflection method.

TABLE 6.7

2–1	2–3	3–2	3–4	4–3	
0·375	0·625	0·610	0·390	0	
0↓	½→	←½	½→	—	
−1372 +139	+1000 +233	+1000 +506	−1830 +324	−1830 0	
0 −76	+203 −127	+116 −74	0 −42	+162 0	
0 +14	−37 +23	−64 +39	0 +25	−21 0	
0 −7	+20 −13	+11 −7	0 −4	+12 0	
−1302	+1302	+1527	−1527	−1679	Uncorrected moments
−272	+272	+318	−318	−350	Corrected moments

6.9 Several Degrees of Freedom

All the examples hitherto considered involving sidesway have had only one possible sway mechanism, so that when the chord rotation of one member is known, those of the other members are uniquely determined. With the hipped portal frame of fig. 6.23(a), however, an infinite number of sway mechanisms

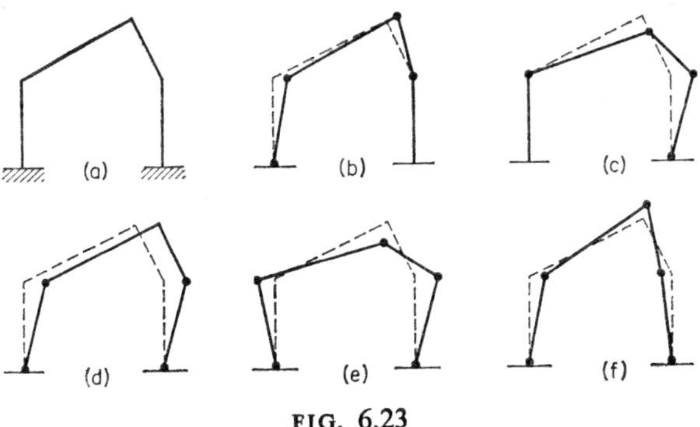

FIG. 6.23

could be formed, of which (b–f) represent five. Not all of these are independent, though: only two are independent and any of the other three can be found from suitable algebraic combinations of them. The deflected shapes in fig. 6.23(d–f), for instance, can be formed by suitable combinations of (b) and (c). Such a frame is said to have two sway degrees of freedom.

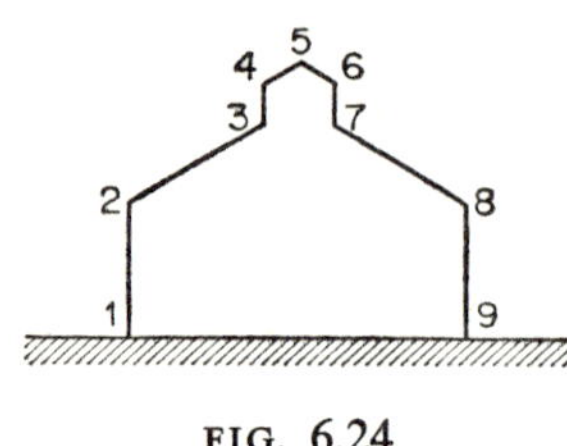

FIG. 6.24

In two dimensions, four bars hinged together form a single-degree-of-freedom mechanism. The frame of fig. 6.22 would form such a mechanism if hinges were put at all the joints, counting the foundation as a fourth bar. The hipped portal of fig. 6.23 would have five hinges and hence one more than would be needed for a single-degree-of-freedom mechanism. The number of degrees of freedom D of a single ring structure of this sort must therefore be

$$D = H - 3$$

where H is the number of effective hinges in the ring. For example, the frame shown in fig. 6.24 will have $9 - 3 = 6$ sway degrees of freedom.

For a two-dimensional multiple ring structure such as those in fig. 6.25, the number of hinges will be

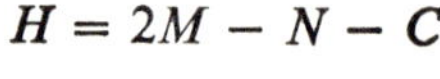

$$H = 2M - N - C$$

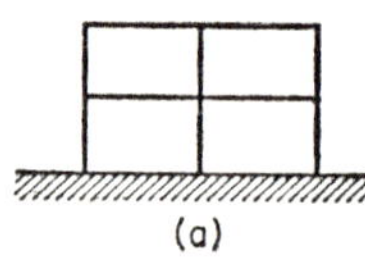

(a)

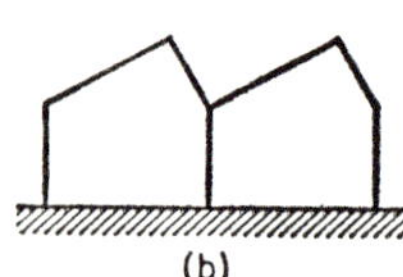

(b)

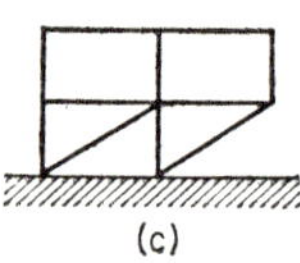

(c)

FIG. 6.25

where M is the total number of members (counting the foundations), N is the number of nodes in the structure and C is the number of additional constraints, such as the fact that the foundation in fig. 6.25(a and b) cannot have a hinge in it at the foot of the centre column. The number of degrees of freedom is then given by

$$D = 2M - N - C - 3R \tag{6.5}$$

where R is the number of rings in the structure. For example, the multiple ring structures of fig. 6.25 have degrees of freedom of

$$D_{(a)} = 2 \times 12 - 9 - 1 - 12 = 2$$
$$D_{(b)} = 2 \times 9 - 8 - 1 - 6 = 3$$
$$D_{(c)} = 2 \times 12 - 8 - 0 - 15 = 1$$

A simpler statement of the number of sway degrees of freedom of a structure is to say that it is equal to the number of restraints which must be inserted into the equivalent mechanism to give a statically determinate structure.

The analysis of a hipped portal frame (fig. 6.26) may be used to illustrate the approach to the solution of structures with several degrees of freedom. The frame is built-in to the foundation at 1 and 5. Each member is uniform with an EI value of 100, while the roof is subjected to a uniform vertical dead load of 1·2/unit length along 2–3 and 3–4. The two sway modes chosen are shown in fig. 6.26.

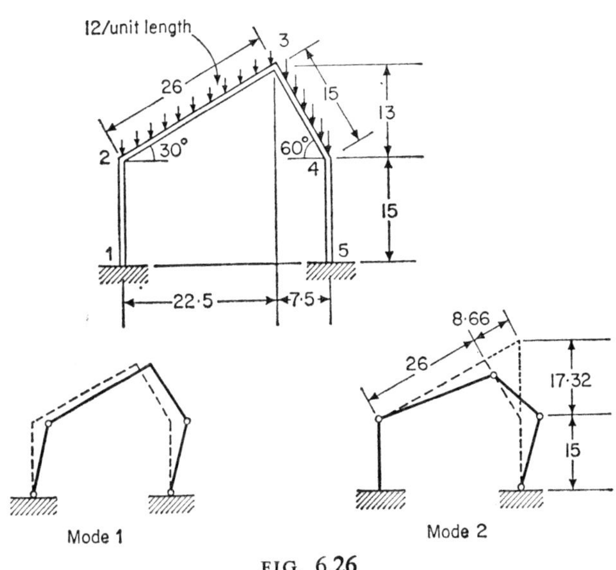

FIG. 6.26

For the no-sway distribution, the fixed-end moments on the roof members are required. The usual formula is used, except that the component of the load perpendicular to the roof must be taken. The fixed-end moments then become

$$M_{23}^F = -\frac{1{\cdot}2 \cos 30 \times 26^2}{12} = -58{\cdot}6 = -M_{32}^F$$

$$M_{34}^F = -\frac{1{\cdot}2 \sin 30 \times 15^2}{12} = -11{\cdot}25 = M_{43}^F$$

The stiffness and distribution factors are

$$S_{21} = \frac{4 \times 100}{15} = 26{\cdot}6 \qquad \therefore\ D_{21} = \frac{26{\cdot}6}{42} = 0{\cdot}634$$

$$S_{23} = S_{32} = \frac{4 \times 100}{26} = 15{\cdot}4 \qquad D_{23} = 0{\cdot}366$$

$$S_{34} = S_{43} = \frac{4 \times 100}{15} = 26{\cdot}6 \qquad D_{32} = 0{\cdot}366$$

$$S_{45} = S_{54} = \frac{4 \times 100}{15} = 26{\cdot}6 \qquad D_{34} = 0{\cdot}634$$

$$D_{43} = D_{45} = 0{\cdot}5$$

All carry-over factors are $\frac{1}{2}$. The no-sway distribution may now be carried out as shown in Table 6.8.

TABLE 6.8

1–2	2–1	2–3	3–2	3–4	4–3	4–5	5–4
0	0·634	0·366	0·366	0·634	0·5	0·5	0
0 0	0 +37·2	−58·6 +21·4	+58·6 −17·3	−11·2 −30·1	+11·2 −5·6	0 −5·6	0 0
+18·6 0	0 +5·5	−8·6 +3·1	+10·7 −2·9	−2·8 −5·0	−15·0 +7·5	0 +7·5	−2·8 0
+2·7 0	0 +0·9	−1·4 +0·5	+1·5 −1·9	+3·7 −3·3	−2·5 +1·2	0 +1·3	+3·7 0
+0·4 0	0 +0·6	−0·9 +0·3	+0·2 −0·3	+0·6 −0·5	−1·6 +0·8	0 +0·8	+0·6 0
+21·7	+44·2	−44·2	+48·6	−48·6	−4·0	+4·0	+1·5

Sway mode 1 moves node points 2 and 4 by an equal horizontal distance. The chord rotations of 1–2 and 3–4 are thus equal, and the initial-sway moments at the top and bottom of each column can be fixed at, say, 1000*f*. The first sway distribution is now carried out in Table 6.9.

TABLE 6.9

1–2	2–1	2–3	3–2	3–4	4–3	4–5	5–4
0	0·634	0·366	0·366	0·634	0·5	0·5	0
+1000 0	+1000 −634	0 −366	0 0	0 0	0 −500	+1000 −500	+1000 0
−317 0	0 0	0 0	−183 +158	−250 +275	0 0	0 0	−250 0
0 0	0 −50	+79 −29	0 0	0 0	+137 −68	0 −69	0 0
−25 0	0 0	0 0	−14 +17	−34 +31	0 0	0 0	−34 0
0 −2	0 −5	+8 −3	0 0	0 0	+15 −7	0 −8	0 −4
+656	+311	−311	−22	+22	−423	+423	+712

The chord rotation relationships involved in mode 2 are rather more difficult to find. The instantaneous centre of rotation of 3–4 is shown at C in fig. 6.26: its distance from points 3 and 4 is also shown. Once this has been found, the chord rotation relationships may be written as

$$\varphi_{23} = -\frac{8{\cdot}66}{26}\varphi_{34} = -0{\cdot}333\varphi_{34}$$

$$\varphi_{45} = -\frac{17{\cdot}32}{15}\varphi_{34} = -1{\cdot}156\varphi_{34}$$

where, as usual, the positive directions of all the chord rotations are taken to be clockwise. Now using the general slope-deflection equation and remembering that at this stage the joint rotations are zero, the initial-sway moment M_{23} may be written

$$M_{23} = (S_{23} + C_{32}S_{32})\varphi_{32} = 23{\cdot}1 \times 0{\cdot}333\varphi_{34}$$

Similarly,

$$M_{34} = -39{\cdot}9\varphi_{34}$$
$$M_{45} = 39{\cdot}9 \times 1{\cdot}156\varphi_{34}$$

If M_{34} is chosen to be 1000, then the other initial-sway moments become

$$M_{23} = -\frac{23{\cdot}1 \times 0{\cdot}333 \times 1000}{39{\cdot}9} = -193$$

$$M_{45} = -\frac{39{\cdot}9 \times 1{\cdot}156 \times 1000}{39{\cdot}9} = -1156$$

The second mode sway distribution is now carried out in Table 6.10.

TABLE 6.10

1–2	2–1	2–3	3–2	3–4	4–3	4–5	5–4
0	0·634	0·366	0·366	0·634	0·5	0·5	0
0	0	−193	−193	+1000	+1000	−1156	−1156
0	+122	+71	−296	−511	+78	+78	0
+61	0	−148	+35	+39	−255	0	+39
0	+94	+54	−27	−47	+127	+128	0
+47	0	−13	+27	+63	−23	0	+64
0	+8	+5	−33	−57	+12	+11	0
+4	0	−16	+2	+6	−28	0	+5
+5	+10	+6	−3	−5	+14	+14	+7
+117	+234	−234	−488	+488	+925	−925	−1041

The next step is to find the factors by which each set of arbitrary sway moments must be multiplied. These are obtained by writing two moment-sway equations, one for each mode. The equation for the first mode is

$$(M_{12} + M_{21})\,\delta\varphi_{21}^{1} + (M_{45} + M_{54})\,\delta\varphi_{45}^{1} = 0$$

where superscripts are used on the virtual displacements to show that they are chord rotations in mode 1. Note that because the roof moves horizontally, the applied load does no work in this mode and so does not come into the equation. For displacement in the second mode, the moment-sway equation is

$$(M_{23} + M_{32})\,\delta\varphi_{23}^{2} + (M_{34} + M_{43})\,\delta\varphi_{34}^{2} + (M_{45} + M_{54})\,\delta\varphi_{45}^{2}$$

$$+26 \times 1{\cdot}2 \times \frac{22{\cdot}5}{2}\,\delta\varphi_{23}^{2} - 15 \times 1{\cdot}2 \times \frac{7{\cdot}5}{2}\,\delta\varphi_{34}^{2} = 0$$

For the purposes of an equilibrium equation a distributed load can, of course, be replaced by an equal concentrated load at its centroid.

Any beam-end moment is now the sum of three terms. For example

$$M_{12} = M_{12}^{0} + f_1 M_{12}^{1} + f_2 M_{12}^{2}$$

Using this and substituting the values from the distribution tables into the moment-sway equations, we get

$$21{\cdot}7 + 44{\cdot}2 + 4{\cdot}0 + 1{\cdot}5 + (656 + 311 + 423 + 712)f_1 + (117 + 234 - 925 - 1041)f_2 = 0$$

$$\begin{aligned} -0{\cdot}333(-44{\cdot}2 + 48{\cdot}6) + (-48{\cdot}6 - 4{\cdot}0) - 1{\cdot}156(4{\cdot}0 + 1{\cdot}5) \\ +[-0{\cdot}333(-311 - 22) + (23 - 423) - 1{\cdot}156(423 + 712)]f_1 \\ +[-0{\cdot}333(-234 - 488) + (488 + 925) - 1{\cdot}156(-925 - 1041)]f_2 \\ -26 \times 1{\cdot}2 \times \frac{22{\cdot}5}{2} \times 0{\cdot}333 - 15 \times 1{\cdot}2 \times \frac{7{\cdot}5}{2} = 0 \end{aligned}$$

These equations may be solved to give the correction factors

$$f_1 = +0{\cdot}0202$$
$$f_2 = +0{\cdot}0706$$

The final moments are found by adding their component terms as set out in Table 6.11.

TABLE 6.11

Moment	M_{12}	M_{21}	M_{32}	M_{43}	M_{54}
No-sway	+21·7	+44·2	+48·6	−4·0	+1·5
Mode 1	+13·2	+6·3	+0·4	−8·5	+14·4
Mode 2	+8·3	+16·5	−34·4	+65·3	−73·5
Total	+43·2	+67·0	+14·6	+52·8	−57·6

The bending moment diagram for the structure is plotted in fig. 6.27.

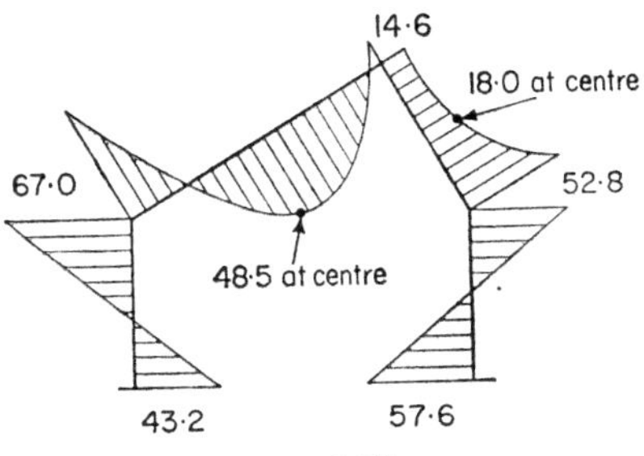

FIG. 6.27

6.10 Multistorey Frames

Multistorey rectangular frames such as the one shown in fig. 6.28 may be handled in the same way as the more general frames dealt with in the previous section, but their regularity of form leads to certain simplifications in the analysis.

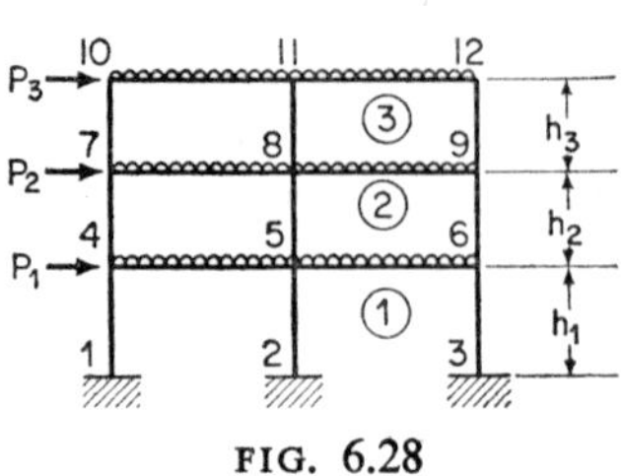

FIG. 6.28

The no-sway distribution for such frames presents no difficulty and is carried out in the usual way. Convergence is usually rapid and it is seldom necessary to carry over the effects of moments at any one floor level further than one storey above and one below. The number of sway degrees of freedom of the structure is equal to the number of stories. The frame in fig. 6.28 has three degrees of freedom, which may be verified by the use of equation (6.5). There are two possibilities for the choice of a rational set of sway mechanisms: those shown at (a) and (b) in fig. 6.29. The second set in which the entire structure above a given storey sways sideways is the better one to use as the number of chord rotations and hence the number of initial sway moments and of terms in the moment-sway equations is less.

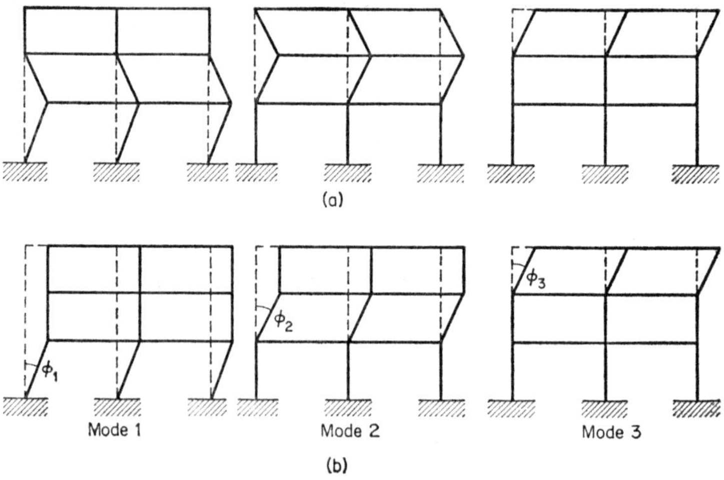

FIG. 6.29 Choices of sway mode

Considering mode 1 in fig. 6.29(b), then by virtual work the moment-sway equation becomes

$$(M_{41} + M_{14} + M_{52} + M_{25} + M_{63} + M_{36})\,\delta\varphi_1 + (P_1 + P_2 + P_3)h_1\,\delta\varphi_1 = 0$$

If we now define a term

$$Q_1 = P_1 + P_2 + P_3$$

where Q_1 is called the *storey shear* and is equal to the total shear force across the first storey, and if we define $\sum M_{c1}$ as the sum of all the moments acting at the tops and bottoms of all columns in the first storey (which may be called the *column sum*), then the moment-sway equation may be rewritten in the form

$$\sum M_{c1} + Q_1 h_1 = 0$$

Similarly, for the other storeys

$$\sum M_{c2} + Q_2 h_2 = 0$$

$$\sum M_{c3} + Q_3 h_3 = 0$$

and for any storey i of a multistorey building

$$\sum M_{ci} + Q_i h_i = 0$$

However, the total moment at the end of a beam is the sum of a no-sway moment and of moments from each sway mode which will have been found as usual in terms of an arbitrary moment multiplied by a correction factor. Thus, for example, moment M_{52} may be written

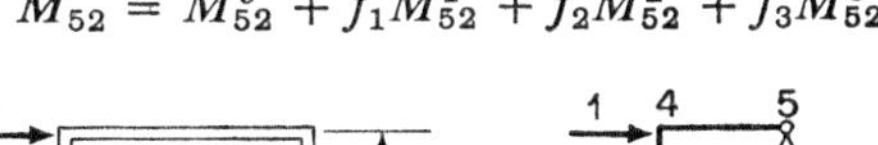

$$M_{52} = M_{52}^0 + f_1 M_{52}^1 + f_2 M_{52}^2 + f_3 M_{52}^3$$

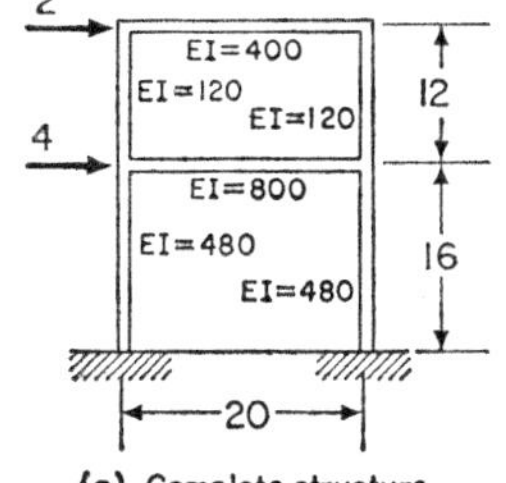

(a) Complete structure

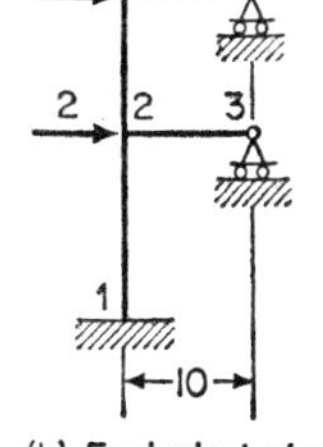

(b) Equivalent structure

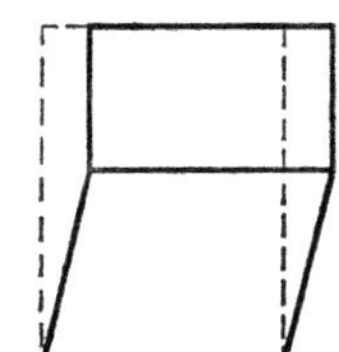

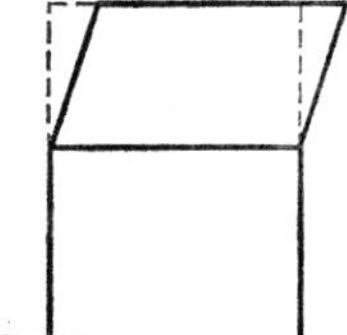

(c) Sway modes

FIG. 6.30

where the superscripts on the moments and the subscripts on the correction factors refer to the various sway modes. The typical moment-sway equation for storey i of an n-storied structure may therefore be written

$$\sum M^0_{ci} + f_1 \sum M^1_{ci} + f_2 \sum M^2_{ci} + \cdots + f_n \sum M^n_{ci} + Q_i h_i = 0 \quad (6.6)$$

For an n-storied structure, n such equations have to be solved for the correction factors $f_1, f_2, \ldots, f_n$.

As an example, consider the frame shown in fig. 6.30(a). The structure is symmetric and the loading on it is antisymmetric so that it can be represented by the equivalent half-frame of fig. 6.30(b). The stiffness and distribution factors are

$$S'_{23} = \frac{3 \times 800}{10} = 240 \qquad \therefore\ D_{23} = 0{\cdot}6$$

$$S_{21} = \frac{4 \times 480}{16} = 120 \qquad D_{21} = 0{\cdot}3$$

$$S_{24} = \frac{4 \times 120}{12} = 40 \qquad D_{24} = 0{\cdot}1$$

$$S_{42} = 40 \qquad D_{42} = 0{\cdot}25$$

$$S'_{45} = \frac{3 \times 400}{10} = 120 \qquad D_{45} = 0{\cdot}75$$

The distribution is now carried out in Table 6.12, in which the initial-sway moments for both modes are arbitrarily taken as 400. The distributed uncorrected moments are obtained in the penultimate line of each column, and these moments are multiplied by the appropriate correction factors to get the final moments in each mode. The moment-sway equations are, for the bottom storey

$$2(M^1_{12} + M^1_{21})f_1 + 2(M^2_{12} + M^2_{21})f_2 + (2 + 4) \times 16 = 0$$

or

$$2(340 + 280)f_1 + 2(-53 - 105)f_2 + 96 = 0$$

and for the top storey

$$2(M^1_{24} + M^1_{42})f_1 + 2(M^2_{24} + M^2_{42})f_2 + 2 \times 12 = 0$$

or

$$2(-40 - 15)f_1 + 2(315 + 285)f_2 + 24 = 0$$

TABLE 6.12

	MODE 1		MODE 2	
Member	4–2	4–5	4–2	4–5
Distribution factor	0·25	0·75	0·25	0·75
Initial-sway moment	0 0	0 0	+400 −100	0 −300
	−20 +5	0 +15	−20 +5	0 +15
	−15	+15	+285	−285
	+1·27	−1·27	−7·89	+7·89

	MODE 1			MODE 2		
Member	2–4	2–1	2–3	2–4	2–1	2–3
Distribution factor	0·10	0·30	0·60	0·10	0·30	0·60
Initial-sway moment	0 −40	+400 −120	0 −240	+400 −40	0 −120	0 −240
	0 0	0 0	0 0	−50 +5	0 +15	0 +30
	−40	+280	−240	+315	−105	−210
	+3·38	−23·7	+20·3	−8·72	+2·91	+5·81

MODE 1	MODE 2
1–2	1–2
0	0
+400 0	0 0
−60 0	−60 0
+340	+7
−28·8	−53
	+1·47

The solution of these equations gives correction factors of

$$f_1 = -0{\cdot}0846$$
$$f_2 = -0{\cdot}0277$$

The total member end moments are obtained in Table 6.13, and the bending moment diagram for the structure is shown in fig. 6.31.

TABLE 6.13

Moment	Mode 1	Mode	Total
M_{42}	+1·27	−7·89	−6·62
M_{24}	+3·38	−8·72	−5·34
M_{21}	−23·7	+2·91	−20·8
M_{23}	+20·3	+5·81	+26·1
M_{12}	−28·8	+1·47	−27·3

The Hardy Cross method of moment distribution which has been described here has one serious disadvantage: that for multistorey structures the sway effects of the various stories must be handled with a separate sway distribution for each storey and with a series of simultaneous equations which must be solved for the storey sway correction factors. For this reason, the technique is seldom used in the analysis of multistorey structures more than a few stories high where sidesway is important—when, for instance, considering wind or seismic loads. In such cases, a computer solution should be used. If, however, a computer is unavailable, then if the structure is a symmetric singlebay structure, the method of cantilever (or Naylor) distribution† may be used in which sidesway is automatically taken into account in the distribution procedure by using suitably modified stiffness and carry-over factors. For general structures, though, the Kani distribution procedure‡ is to be recommended in which storey sways as well as joint rotations are systematically

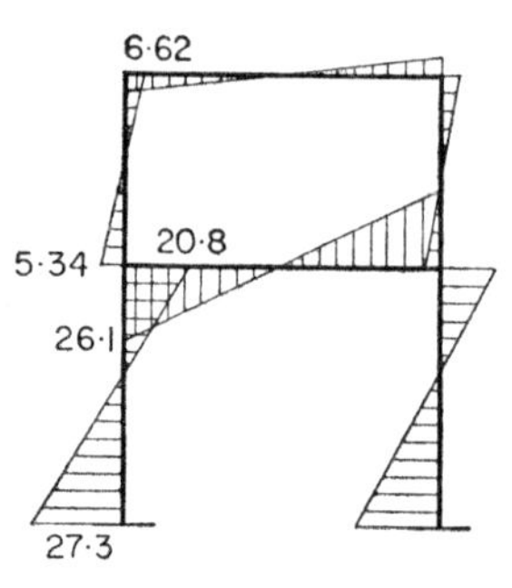

FIG. 6.31

† Naylor, N., 'Sidesway in Building Frames', *Structural Engineer,* April 1950; also summarised in Gray, C. S. *et al., Steel Designer's Manual,* Lockwood, London, 1956, pp. 201–15.

‡ Kani, G., *Analysis of Multi-Storey Frames,* Lockwood, London, 1957.

relaxed. The Kani technique has two added advantages over the Hardy Cross method. Firstly, the iteration is carried out on the total beam-end moments rather than on the out-of-balance moments so that the further the iteration is carried the more accurate are the results; and secondly, any error introduced in the iteration process is automatically removed by subsequent cycles of iteration. However, the Kani procedure is rather more complicated than the Cross method and space does not allow its inclusion in this book.

PROBLEMS

6.1 Find the bending moments at points B and C of the uniform beam shown in fig. 6.32.

Answer: 11·4, 13·9

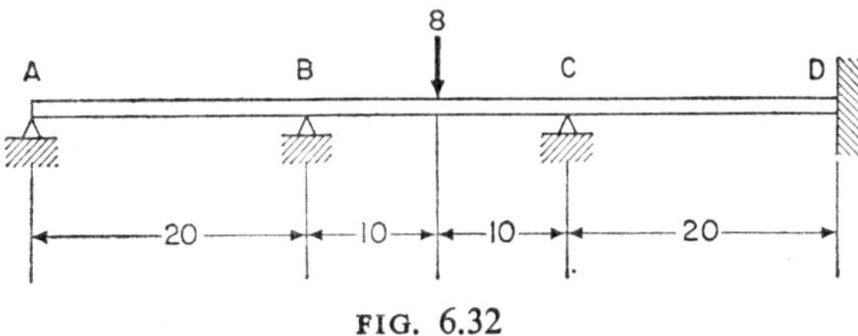

FIG. 6.32

6.2 Determine the reactions of the plane frame of fig. 6.33. Neglect the change in length of the members. All members are uniform and have the same section stiffness.

Answer: 3·89, 3·47, 1·94

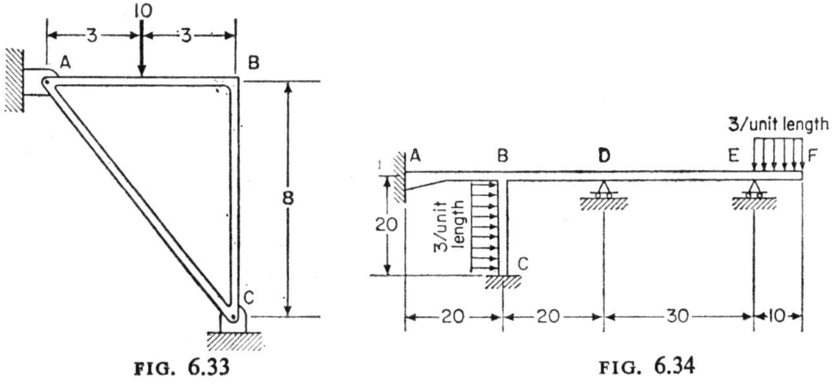

FIG. 6.33 FIG. 6.34

6.3 Fig. 6.34 shows a rigid frame. All members have a uniform section stiffness EI, except that member AB has a straight haunch as shown. The

PCA Handbook of Frame Constants gives the following information for member AB:

$$S_{AB} = \frac{8{\cdot}29EI}{L}; \qquad C_{AB} = 0{\cdot}449$$

$$S_{BA} = \frac{4{\cdot}71EI}{L}; \qquad C_{BA} = 0{\cdot}791$$

Draw the bending moment diagram for the frame, plotting bending moments on the tension sides of all members.

Answer: Moments at B = 75·2, 46·2, 29·2

6.4 The five-span bridge shown diagrammatically in fig. 6.35 is constructed of uniform members whose moments of inertia are I for horizontal members and $2I$ for vertical members. The structure is connected to its piers and its abutments by hinges at A, H, I, J, K, and L, and there are

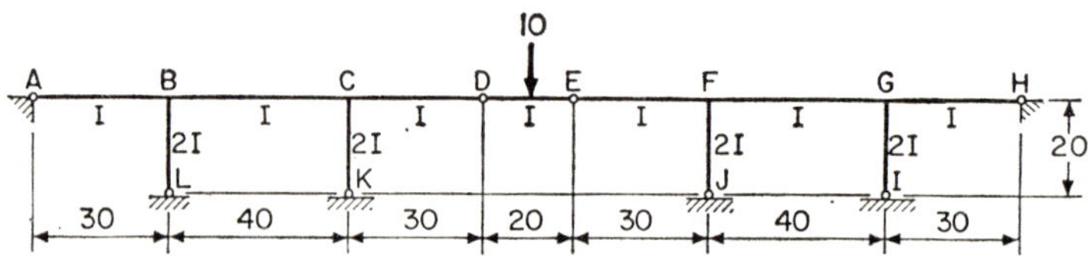

FIG. 6.35

also hinges at points D and E. Find the bending moments at B and C due to a concentrated load of 10 applied in the centre of the bridge. Sketch the bending moment diagram for the complete structure indicating values at important points.

6.5 The highway bridge shown in fig. 6.36 carries a uniformly distributed load of 1·0/unit length over its entire length. The section stiffness of the

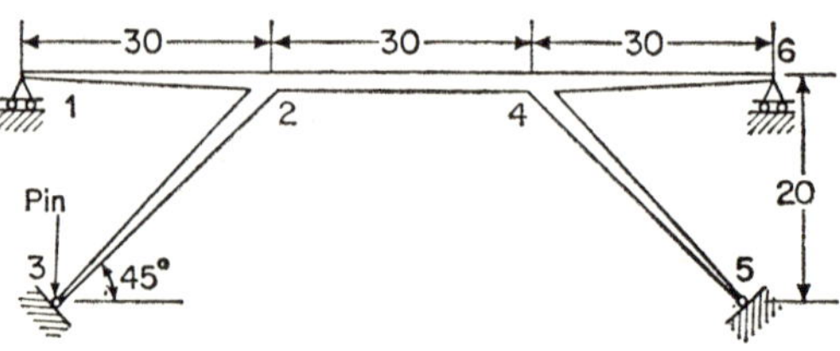

FIG. 6.36

prismatic member 2–4 is EI, and the other members are tapered and are of the same shape, so that the unmodified stiffness and carry-over factors are given by:

$$S_{12}L_{12} = S_{32}L_{23} = 2{\cdot}4EI$$
$$S_{21}L_{12} = S_{23}L_{23} = 4{\cdot}8EI$$
$$C_{12} = C_{32} = 0{\cdot}70$$
$$C_{21} = C_{23} = 0{\cdot}35$$

The fixed-end moments for a uniformly distributed load applied to a member of the same shape as 1–2 are:

$$M_{12}^F = 0{\cdot}060wL^2$$
$$M_{21}^F = 0{\cdot}100wL^2$$

Find the *vertical* component of reaction at support 3.

Answer: 33·6

6.6 Analyse the structure of fig. 6.37 and sketch the final bending moment diagram.

Answer: Moments at 2 = 17·6, 14·0, 3·5

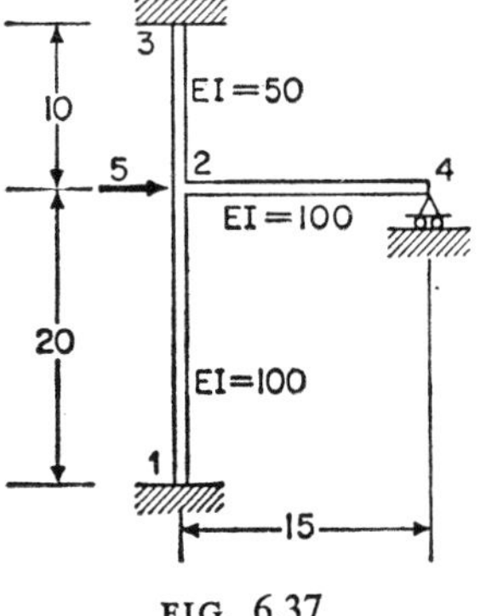

FIG. 6.37

6.7 A loaded steel frame shown in fig. 6.38 is supported at B by the free end of a steel cantilever of length 10. Draw the bending moment diagram for the frame and sketch its deflected shape. Also compute the lateral

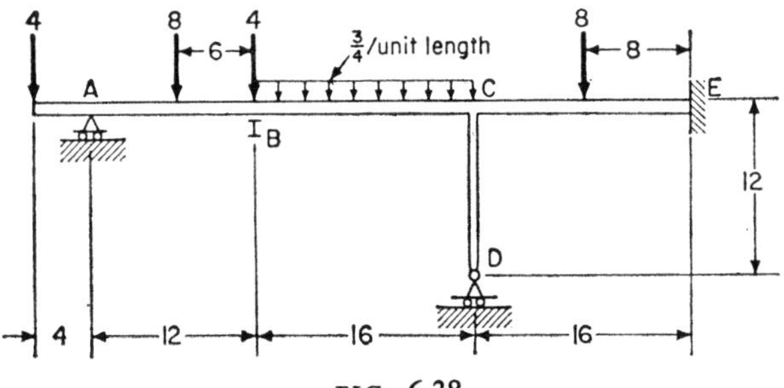

FIG. 6.38

displacement of the column base. $EI = 3 \times 10^4$ for all members of the frame and for the cantilever.

Answer: 0·0412

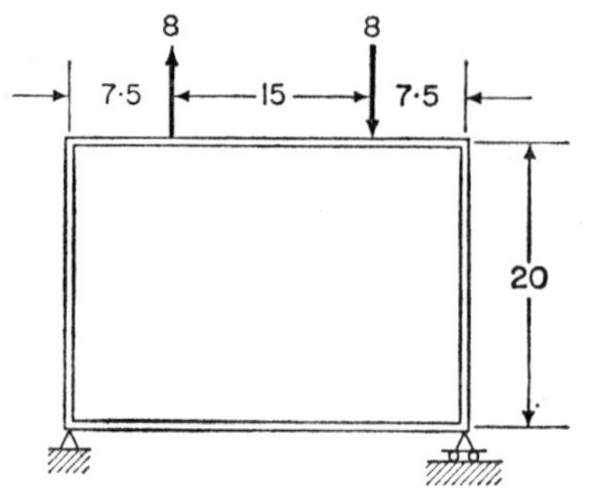

FIG. 6.39

6.8 The Vierendeel truss shown in fig. 6.39 is acted on by two loads of 8 units as shown. All members have the same uniform cross-section.

(a) Find the bending moments at the corners.

(b) Sketch the bending moment diagram.

Answer: 3·7, 3·8

6.9 A continuous uniform beam ABCD is built-in at A and D and has three uniform spans of length 30. At C the beam is simply supported, but at B the beam rests on the centre of a simply supported transverse beam of length 30 and with the same cross-section as ABCD. A uniformly distributed load of 1·2/unit length is applied over the whole length of the beam ABCD. Find the support moments at A, B, C and D, and sketch the bending moment diagram.

Answer: 180, 6, 142, 65

6.10 A highway bridge has the same dimensions and support conditions as the bridge in fig. 6.36, but all its members are uniform, with section stiffnesses *EI* except for the central span 2–4, which has a section stiffness 2*EI*. The bridge is loaded with a concentrated force of 20 applied vertically at point 2. Find the moments at 2, and hence sketch the bending moment diagram for the bridge.

Answer: 39, 52, 91

6.11 All members of the frame shown in fig. 6.40 have the same length and section stiffness.

(a) Find the joint bending moments.

(b) Sketch the bending moment diagram.

Answer: 49·2, 49·0, 23·6

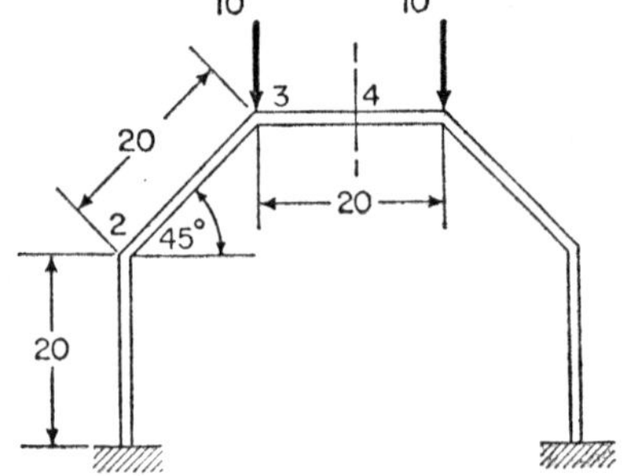

FIG. 6.40

6.12 The hipped portal frame of fig. 6.41 is formed of prismatic members with the same section properties. Find the moments due to a wind loading of 150/unit length pressure and suction on members 1–2 and 4–5, 30/unit length pressure on 2–3 and 135/unit length suction on 3–4.

Answer: $M_{21} = 38{,}600$

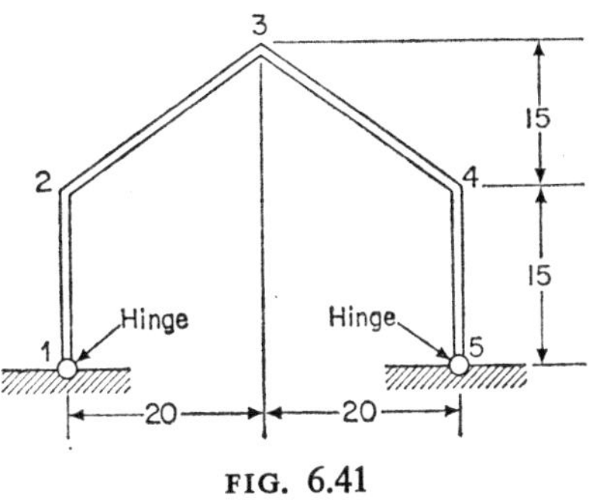

FIG. 6.41

6.13 A one-bay three-storey frame structure has a span of 20, and each storey has a height of 10. Both columns are built-in to the foundation. The structure is subjected to an inward uniformly distributed load of 100/unit length acting over the whole height (30) of one of the columns, and a concentrated lateral load of 1000 acting at the top of the structure. The section stiffness of the columns is EI, and that of the beams is $2EI$. Find the moments in the structure and sketch the bending moment diagram.

Answer: Moment at top of ground floor column = 5740

7

Work and Energy Methods

7.1 Introduction

The previous two chapters dealt with force and displacement methods of structural analysis from a direct point of view: that is, force–displacement relationships were formulated for the various elements of a structure, these were assembled to give the force–displacement characteristics of the structure as a whole, and compatibility or equilibrium equations were derived whose solution provided the required results. The same results could, however, have been obtained less directly using work or energy methods. The study of such indirect methods is the subject of this chapter.

The various work and energy approaches can still be classified under the two categories of force and displacement methods. If forces are chosen as unknowns, they are still found by solving compatibility equations; and unknown displacements require the formulation of equilibrium equations. The main difference between direct and indirect approaches is that the latter are expressed in more general terms: they involve the calculation of the strain energy (or rather, its derivatives) of the entire structure or the work done on or within it. Because of this increased generality, it is natural that work and energy methods should find their greatest use in the analysis of more complex structures and in particular, continuum problems. They are powerful methods, but their very power and generality lead to the disadvantage that for many structural problems they are more cumbersome and complicated to use than direct methods.

This chapter does not attempt to give a full and rigorous coverage to work and energy methods: it merely summarises and illustrates the more important techniques. A more thorough treatment of the subject is given in the two references at the end of the chapter. The first, by Argyris and Kelsey, is terse and rigorous. It appears to be the first attempt to bring together the various types of energy analysis and treat them as a coordinate whole; and for this reason it has been the basis of much subsequent work in the fields both of work and energy methods and of matrix analysis. The second reference, by Neal, is more readable and is thoroughly recommended as background reading in the subject.

7.2 Strain Energy and Complementary Energy

The strain energy in a body is the potential energy stored in the body by reason of its elastic deformation. All parts of such a body will store strain energy if they are subjected to strain though in general the amount of strain energy per unit volume will vary from point to point. This quantity, called *strain energy density*, may be defined as

$$\bar{U} = \int_0^{\varepsilon} [\sigma_{xx}\,d\varepsilon_{xx} + \sigma_{yy}\,d\varepsilon_{yy} + \sigma_{zz}\,d\varepsilon_{zz} + \sigma_{xy}\,d\varepsilon_{xy} + \sigma_{yz}\,d\varepsilon_{yz} + \sigma_{zx}\,d\varepsilon_{zx}]$$

where σ_{xx} is the direct stress in the x-direction, σ_{xy} is the shear stress on the x-face in the y-direction and ε_{xx} and ε_{xy} are the equivalent direct and shear strains. The strain energy density $\bar{U}$ is thus equal to the integral of each stress at a point with regard to the strain from zero to the final strain: in fig. 7.1(a)

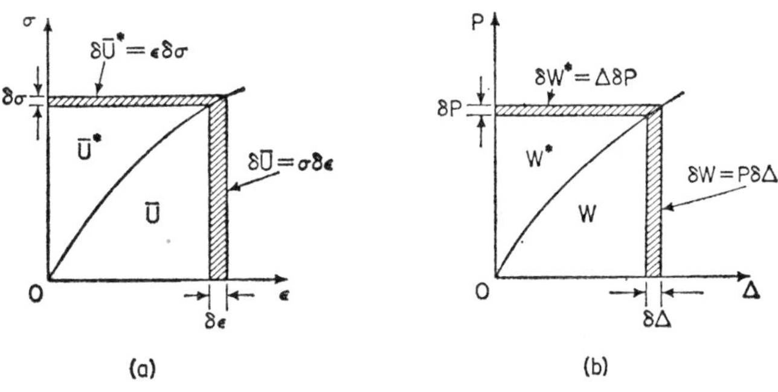

FIG. 7.1

(which is drawn for the one-dimensional situation of one stress and one strain only) $\bar{U}$ is represented by the area under the curve. In this case,

$$\bar{U} = \int_0^{\varepsilon} \sigma\,d\varepsilon \tag{7.1}$$

For a linearly elastic material, the integration could be carried out directly and the strain energy density written as

$$\bar{U} = \tfrac{1}{2}[\sigma_{xx}\varepsilon_{xx} + \sigma_{yy}\varepsilon_{yy} + \sigma_{zz}\varepsilon_{zz} + \sigma_{xy}\varepsilon_{xy} + \sigma_{yz}\varepsilon_{yz} + \sigma_{zx}\varepsilon_{zx}]$$

The total strain energy U in a body will be the integral of the strain energy density over the total volume of the body, or

$$U = \int_V \bar{U}\,dV$$

For an elastic body the total work W done on the body by external forces will be equal to the strain energy (apart from self-strain effects), or

$$U = W$$

For a single load P and displacement Δ, W is the area beneath the curve in fig. 7.1(b) which is the integral of P with regard to Δ, or

$$W = \int_0^\Delta P \, \mathrm{d}\Delta \tag{7.2}$$

The areas above the two curves in fig. 7.1 are the complementary energy density and complementary work, defined as

$$\bar{U}^* = \int_0^\sigma \varepsilon \, \mathrm{d}\sigma \tag{7.3}$$

and

$$W^* = \int_0^P \Delta \, \mathrm{d}P \tag{7.4}$$

These quantities have no particular physical significance: they are merely mathematical quantities defined by equations (7.3) and (7.4); but as such they are helpful in deriving various results in a general way to include the effect of non-linear elasticity. For linearly elastic bodies, of course,

$$\bar{U}^* = \bar{U}$$

and

$$W^* = W$$

7.3 The Principle of Virtual Displacements

It can be seen from fig. 7.1 that for some given load P and displacement Δ, if the displacement is given a small variation $\delta\Delta$, then the work W will vary by an amount

$$\delta W = P \, \delta\Delta \tag{7.5}$$

Similarly, a small variation in strain $\delta\varepsilon$ will cause a variation in strain energy density of

$$\delta\bar{U} = \sigma \, \delta\varepsilon \tag{7.6}$$

The *principle of virtual displacements* or virtual work says that, if a body is in internal and external equilibrium, then

$$\delta W = \delta U \tag{7.7}$$

This rather bald statement of the principle may be amplified by saying that:

> If a body acted upon by some external force system is in a state of equilibrium, then if any small (virtual, or imaginary) displacement system is applied to it which satisfies both the conditions of compatibility and the boundary conditions of the body, then the work done by the external forces acting through the virtual external displacements is equal to the change in strain energy due to the stresses within the body acting through the virtual strains.

For a rigid body there can be no strain energy or internal work, so that equation (7.7) reduces to

$$\delta W = 0 \tag{7.8}$$

This form of the principle of virtual work was discussed in Section 1.1.2; and the examples given there should be compared with those given subsequently in the present section.

An alternative statement of the principle is that, if the external and internal forces are P and F, and if the external and internal displacements are Δ and u, then

$$\begin{aligned}\sum P\,\delta\Delta &= \sum F\,\delta u \\ &= \int_V \varepsilon\,\delta\sigma\,\mathrm{d}V \end{aligned} \tag{7.9}$$

A general proof of the principle of virtual displacements is rather complicated and will not be given here—such a proof is given by Argyris in reference 1. Instead, the principle will be shown to be true in various specific cases.

Consider, for instance, the bar shown in fig. 7.2. The external force system acting on the bar is a tensile force P. The internal force system within the bar resulting from this is a constant uniaxial stress of value $\sigma = P/A$. We can now choose a virtual displacement system which need be subject only to the restriction that the compatibility conditions of the bar are fulfilled—meaning that no gaps should occur in the bar. A suitable choice of displacement system would be a constant small strain $\delta\varepsilon$ throughout the bar in the x-direction. The external displacement compatible with this is an elongation of the bar by an amount $\delta\Delta = \delta\varepsilon L$. The external work is thus

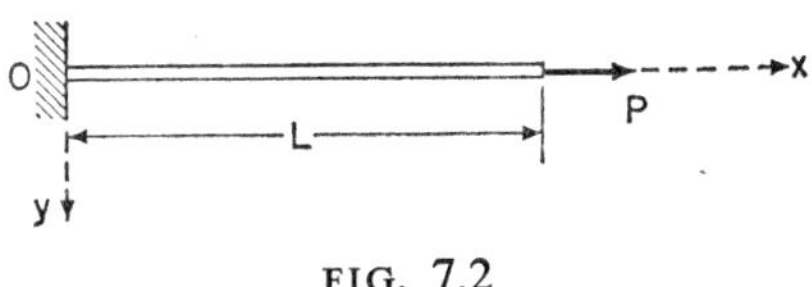

FIG. 7.2

$$\delta W = P\,\delta\varepsilon L$$

and the strain energy is

$$\delta U = \int_0^L \frac{P}{A} A \, \delta\varepsilon \, dx = P \, \delta\varepsilon L$$

Hence

$$\delta U = \delta W$$

and the principle has been shown to hold. In fact it would still hold no matter what the choice of displacement system, for if instead of a constant strain, any arbitrary variation of strain $\delta\varepsilon(x)$ were chosen as the internal virtual displacement system then the external displacement would be an axial extension of $\int_0^L \delta\varepsilon(x) \, dx$. Hence the external work done would be

$$\delta W = P \int_0^L \delta\varepsilon(x) \, dx$$

The change in strain energy would be

$$\delta U = \int_0^L \frac{P}{A} A \, \delta\varepsilon(x) \, dx = P \int_0^L \delta\varepsilon(x) \, dx$$

and

$$\delta U = \delta W$$

as before.

As a slightly less elementary example, consider the pin-jointed truss of fig. 7.3. If a vertical external force P is applied at 5 as shown, then the corresponding internal bar forces can easily be found by the use of statics when the structure is in equilibrium: the bar forces in members 1–2, 1–4 and 3–4 are, for instance

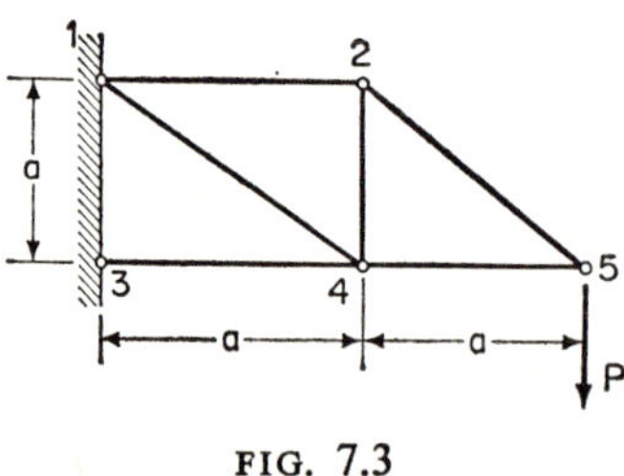

FIG. 7.3

$$F_{12} = P$$
$$F_{14} = \sqrt{2}P$$
$$F_{34} = -2P$$

If the system of virtual internal displacements is chosen quite arbitrarily as a constant tensile strain $\delta\varepsilon$ in member 1–4 with zero strain in all other members, then the bar elongation of 1–4 will be $\sqrt{2}a \, \delta\varepsilon$ and the resulting external displacement of 5 will be $2a \, \delta\varepsilon$ downwards. The external work done by the force system acting through the displacement system will therefore be

$$\delta W = P . 2a \, \delta\varepsilon$$

and the internal work will be

$$\delta U = F_{14}.\sqrt{2}a\,\delta\varepsilon$$
$$= \sqrt{2}P.\sqrt{2}a\,\delta\varepsilon$$

Thus once again equation (7.7) is shown to hold.

It must be emphasised that the displacement system is not caused by the force system and is quite independent of it. It could be thought of as being produced by some other force system; but this is unnecessary. After all, the displacement system is never actually applied to the physical structure—it is just a convenient concept.

As the principle of virtual displacements only holds if the internal and external forces on a body are in equilibrium, the principle is itself a statement of equilibrium. It can therefore be used to solve problems of statics; which it does, however, by turning them into problems of geometry. For instance, suppose it is required to find the force F_{34} in member 3–4 of the truss shown in fig. 7.3 due to a vertical force P at 5. This could be done by applying a small extension δu in member 3–4, which consideration of the kinematics of the system shows would cause point 5 to rise by an amount of $2\delta u$. For equilibrium

$$\delta U = \delta W$$

so

$$F_{34}.\delta u = -P.2\delta u \tag{7.10}$$

and

$$F_{34} = -2P$$

The same result could have been achieved if instead of the small internal displacement δu, an extension of unity had been applied to member 3–4; but in this case the formulation would have been somewhat simpler as equation (7.10) would have become

$$F_{34}.1 = -P.2$$

and F_{34} could have been obtained directly. This modification of the principle of virtual displacements is known as the *unit-displacement method*. In general, if some internal force F is required in a member, then a unit displacement (extension, for a truss) is applied to the member. This will cause a change in the geometry of the structure and some point i will receive a displacement u_i due to the unit displacement. Then if P_i is the external load at i, the required force F is given by

$$F = \sum_i P_i u_i \tag{7.11}$$

where the summation is taken over all the external loads and their corresponding displacements.

For a problem involving statics, it is generally preferable to use the principle of virtual displacements rather than equations of equilibrium only if the geometry of the structure is more easily handled than its statics.

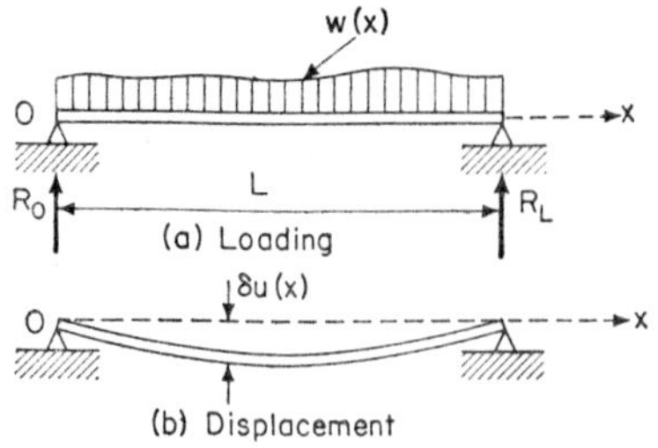

(a) Loading

(b) Displacement

FIG. 7.4

The principle works equally well for beams and frame structures. Consider, for example, a simply supported beam (fig. 7.4) subjected to some arbitrarily distributed continuous load $w(x)$ per unit length, which produces end reactions R_0 and R_L. The bending moment at any point in the beam will be

$$M(x) = \iint w(x)\,\mathrm{d}x\,\mathrm{d}x + R_0 x \tag{7.12}$$

The beam is now given a virtual displacement of $u(x)$ along its length: $u(x)$ is a continuous function satisfying the requirement of compatibility and it is assumed to be zero at the ends of the beam to fulfil the boundary conditions; but apart from these restrictions the value of $u(x)$ is quite arbitrary. The curvature at any point will be $\mathrm{d}^2u(x)/\mathrm{d}x^2$. The external work done by the load $w(x)$ acting through the displacement is

$$\delta W = \int_0^L w(x)u(x)\,\mathrm{d}x$$

and integrating by parts

$$\delta W = \left[\left[\int_0^L w(x)\,\mathrm{d}x\,.\,u(x)\right]\right|_0^L - \int_0^L \frac{\mathrm{d}u(x)}{\mathrm{d}x}\left(\int_0^L w(x)\,\mathrm{d}x\right)\mathrm{d}x$$

$$= -\int_0^L \frac{\mathrm{d}u(x)}{\mathrm{d}x}\left(\int_0^L w(x)\,\mathrm{d}x\right)\mathrm{d}x \tag{7.13}$$

as $u(0) = U(L) = 0$. The internal work done is

$$\delta U = \int_0^L M(x)\frac{\mathrm{d}^2u(x)}{\mathrm{d}x^2}\,\mathrm{d}x$$

which becomes, again integrating by parts,

$$\delta U = \left[M(x)\frac{\mathrm{d}u(x)}{\mathrm{d}x}\right]\Bigg|_0^L - \int_0^L \frac{\mathrm{d}M(x)}{\mathrm{d}x}\cdot\frac{\mathrm{d}u(x)}{\mathrm{d}x}\,\mathrm{d}x$$

From (7.11), remembering that $M(0) = M(L) = 0$,

$$\delta U = -\int_0^L \left(\int_0^L w(x)\,\mathrm{d}x\right) \frac{\mathrm{d}u(x)}{\mathrm{d}x}\,\mathrm{d}x$$

$$= \delta W$$

and the principle has been demonstrated.

If the principle of virtual displacements may be used to set up equilibrium equations then it could evidently be used to form the equations necessary to analyse statically indeterminate structures by the displacement method. In fact, it is seldom used in its general form for the solution of frame structures as it is more cumbersome than the direct method, although it was used in its rigid-body form (equation (7.8)) in Chapter 6 to set up moment-sway equilibrium equations. However, as a general illustration of its use (in the unit-displacement form) the pin-jointed truss of fig. 7.5 will now be analysed: all members are made of the same material and have the same cross-sectional area except for member 0–1 whose area is twice that of the others.

As for other forms of the displacement method of analysis, the first step is to choose a suitable number of unknown displacements. In this case the truss is 2-times kinematically indeterminate; and the obvious choice of unknowns is to take the horizontal and vertical displacements Δx and Δy of point 0. It must be emphasised that these are not virtual but real displacements which actually exist in the frame and which are caused by the application of the load P.

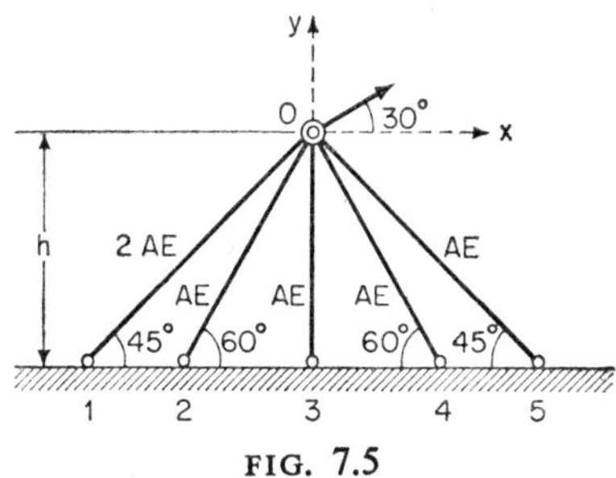

FIG. 7.5

The calculations are set out in Table 7.1. First, the elongations of the members are calculated in columns 3 and 4 due to unit displacements applied at the apex in the x and y directions: the elongations due to Δ_x and Δ_y are evidently these quantities multiplied by Δ_x and Δ_y. Next, the member forces due to Δ_x and Δ_y are calculated by multiplying the elongations by AE/L for each member. Up to this stage the calculations are identical to those which would have had to be performed if a direct stiffness-method analysis were being carried out; for this method, too, would have required the member forces in terms of the unknown displacements. However, whereas the direct approach would continue by using statics to obtain the necessary two equilibrium equations in terms of Δ_x and Δ_y, the unit-displacement method derives what are ultimately the same equations by calculating the work done

TABLE 7.1

1	2	3	4	5	6	7	8
Member	Length	Elongation due to unit displacement in direction		Member forces due to		Work done due to unit displacement in direction	
		x	y	Δx	Δy	x	y
0–1	$\sqrt{2}h$	$\frac{\sqrt{2}}{2}$	$\frac{\sqrt{2}}{2}$	$\frac{2AE}{\sqrt{2}h}\cdot\frac{\sqrt{2}}{2}\Delta_x$	$\frac{2AE}{\sqrt{2}h}\cdot\frac{\sqrt{2}}{2}\Delta_y$	$\frac{2AE}{2h}\cdot\left(\frac{\sqrt{2}}{2}\Delta_x + \frac{\sqrt{2}}{2}\Delta_y\right)$	$\frac{2AE}{2h}\left(\frac{\sqrt{2}}{2}\Delta_x + \frac{\sqrt{2}}{2}\Delta_y\right)$
0–2	$\frac{2}{\sqrt{3}}h$	$\frac{1}{2}$	$\frac{\sqrt{3}}{2}$	$\frac{\sqrt{3}AE}{2h}\cdot\frac{1}{2}\Delta_x$	$\frac{\sqrt{3}AE}{2h}\cdot\frac{\sqrt{3}}{2}\Delta_y$	$\frac{\sqrt{3}AE}{4h}\left(\frac{1}{2}\Delta_x + \frac{\sqrt{3}}{2}\Delta_y\right)$	$\frac{3AE}{4h}\left(\frac{1}{2}\Delta_x + \frac{\sqrt{3}}{2}\Delta_y\right)$
0–3	h	0	1	0	$\frac{AE}{h}\cdot\Delta_y$	0	$\frac{AE}{h}\cdot\Delta_y$
0–4	$\frac{2}{\sqrt{3}}h$	$-\frac{1}{2}$	$\frac{\sqrt{3}}{2}$	$-\frac{\sqrt{3}AE}{2h}\cdot\frac{1}{2}\Delta_x$	$+\frac{\sqrt{3}AE}{2h}\cdot\frac{\sqrt{3}}{2}\Delta_y$	$\frac{\sqrt{3}AE}{4h}\left(\frac{1}{2}\Delta_x - \frac{\sqrt{3}}{2}\Delta_y\right)$	$\frac{3AE}{4h}\left(-\frac{1}{2}\Delta_x + \frac{\sqrt{3}}{2}\Delta_y\right)$
0–5	$\sqrt{2}h$	$-\frac{\sqrt{2}}{2}$	$\frac{\sqrt{2}}{2}$	$-\frac{AE}{\sqrt{2}h}\cdot\frac{\sqrt{2}}{2}\Delta_x$	$\frac{AE}{\sqrt{2}h}\cdot\frac{\sqrt{2}}{2}\Delta_y$	$\frac{AE}{2h}\left(\frac{\sqrt{2}}{2}\Delta_x - \frac{\sqrt{2}}{2}\Delta_y\right)$	$\frac{AE}{2h}\left(-\frac{\sqrt{2}}{2}\Delta_x + \frac{\sqrt{2}}{2}\Delta_y\right)$
					TOTAL	$\frac{AE}{h}(1{\cdot}493\Delta_x + 0{\cdot}353\Delta_y)$ $=\delta U_x$	$\frac{AE}{h}(0{\cdot}353\Delta_x + 2{\cdot}360\Delta_y)$ $=\delta U_y$

by the member forces when unit displacements are applied to 0 in the x and y directions: these quantities are the sums of the various terms in columns 7 and 8. The values of external work done by the load P when unit displacements are applied in the two directions are

$$\delta W_x = \frac{\sqrt{3}}{2} P.1$$

and

$$\delta W_y = \frac{1}{2} P.1$$

The internal and external work done may now be equated to give

$$\frac{AE}{h}(1{\cdot}493\Delta_x + 0{\cdot}353\Delta_y) = \frac{\sqrt{3}P}{2}$$

$$\frac{AE}{h}(0{\cdot}353\Delta_x + 2{\cdot}360\Delta_y) = \frac{P}{2}$$

whence the unknown displacements are found to be

$$\Delta_x = 0{\cdot}550\frac{Ph}{AE}$$

$$\Delta_y = 0{\cdot}130\frac{Ph}{AE}$$

Back-substitution into columns 5 and 6 of Table 7.1 now gives the member forces: for instance

$$F_{04} = \left(-\frac{\sqrt{3}}{4}\cdot\frac{AE}{h} \times 0{\cdot}550 + \frac{3}{4}\cdot\frac{AE}{h} \times 0{\cdot}130\right)\frac{Ph}{AE}$$

$$= -0{\cdot}140P$$

and so on for the forces in the other four members.

7.4 Castigliano's Theorem, Part I

Fig. 7.6 represents a typical elastic structure which has three displacements Δ_1–Δ_3 applied to it: these displacements cause reactive forces P_1–P_3 to occur at the same points (such a situation may be imagined to occur in a displacement-method analysis when the unknown displacements are applied to the structure), and due to the deformations the structure will contain strain energy U. If one of the displacements, at point 2, say, is now given a small variation $\delta\Delta_2$, then the work done on the structure will be

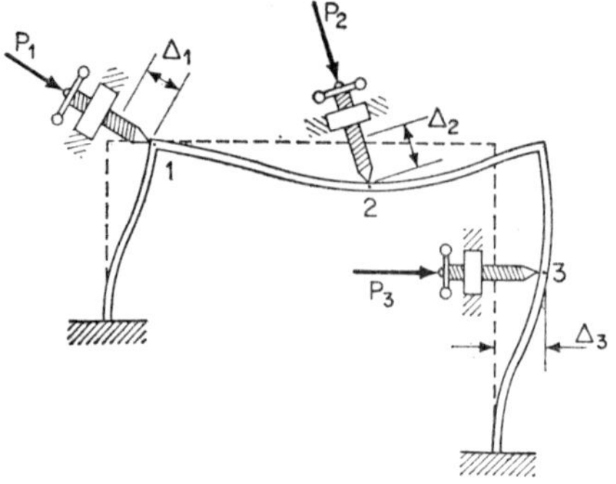

FIG. 7.6

$$\delta W = P_2\,\delta\Delta_2$$

and as we know from the principle of virtual displacements that

$$\delta W = \delta U$$

then the change in strain energy will be

$$\delta U = P_2\,\delta\Delta_2$$

and in the limit as $\delta\Delta_2 \to 0$, we can write

$$P_2 = \frac{\partial U}{\partial \Delta_2} \tag{7.14}$$

if the strain energy is expressed as a function of the applied displacements. This may be stated:

> The force applied to a body at some point is equal to the partial derivative of the strain energy of the body with regard to the displacement at that point.

This is *Castigliano's theorem, Part I.* It should be noted that the theorem is not restricted to linear elasticity.

This theorem can be used instead of the unit displacement method. Its proof shows it to be merely an alternative statement of the principle of virtual displacements, but as it uses strain energy (in terms of displacements) and because this (or its derivative) has to be evaluated, it is somewhat more awkward to use.

The calculations which would be involved in the analysis of the truss shown in fig. 7.5 are similar to those given in Table 7.1 for the unit displacement method. Consider member 0–2 for example. Its extension will be

$$\Delta_{02} = \frac{\Delta_x}{2} + \frac{\sqrt{3}}{2}\Delta_y$$

and as it is linearly elastic and its length is $2h/\sqrt{3}$ its strain energy in terms of its extension will be

$$U_{02} = \frac{1}{2}\Delta_{02}^2 \, AE \frac{\sqrt{3}}{2h}$$

$$= \frac{AE\sqrt{3}}{4h}\left(\frac{\Delta_x}{2} + \frac{\sqrt{3}}{2}\Delta_y\right)^2 \tag{7.15}$$

Hence by Castigliano's Theorem, Part I, the external force at 0 in direction x due to the strain energy in 0–2 will be

$$P_x = \frac{\partial U_{02}}{\partial \Delta_x} = \frac{2AE\sqrt{3}}{4h} \cdot \frac{1}{2} \cdot \left(\frac{\Delta_x}{2} + \frac{\sqrt{3}}{2}\Delta_y\right)$$

which is equal to the equivalent term in column 7 of Table 7.1. It is not necessary, however, to follow equation (7.15) and form the strain energy directly, for all that is required is its derivative. Thus we could have written instead

$$\frac{\partial U_{02}}{\partial \Delta_x} = \frac{AE\sqrt{3}}{2h} \cdot \Delta_{02} \cdot \frac{\partial \Delta_{02}}{\partial \Delta_x}$$

It will be seen that the derivative on the right-hand side of this equation is equal to the extension of the member due to a unit displacement at 0 in the x-direction, given in column 3 of Table 7.1, while the rest of the term represents the member forces given in columns 5 and 6.

Therefore, for a linearly elastic structure, an analysis using Castigliano's theorem, Part I, is exactly the same computationally as one employing the unit-displacement method. The converse is not necessarily true, in that the unit-displacement method has the added generality that any choice of unit-displacement system may be made provided that the boundary and compatibility conditions are fulfilled, whereas only the actual unknown displacements may be used if Castigliano's theorem is employed.

7.5 The Principle of Virtual Forces

Returning now to fig. 7.1(b), it can be seen that for a given force/displacement situation, a small change in force δP causes a change in the complementary work of a structure of

$$\delta W^* = \Delta \, \delta P$$

Similarly, a small variation $\delta\sigma$ in stress at some point will cause a change in complementary energy density of

$$\delta\bar{U}^* = \varepsilon\,\delta\sigma$$

The complementary work and energy may be connected by the *principle of virtual forces* which states that

> If an elastic structure subjected to a system of strains and displacements fulfils its kinematic boundary and compatibility conditions, then if a small (virtual or imaginary) set of forces is applied to the structure, the complementary work done by the external virtual forces acting through the displacements of the structure is equal to the complementary energy change within the structure.

That is,

$$\delta W^* = \delta U^* \tag{7.16}$$

An alternative statement of the principle is that, if the external and internal forces are P and F respectively, and if the external and internal displacements are Δ and u, then

$$\begin{aligned} \sum \Delta\,\delta P &= \sum u\,\delta F \\ &= \int_v \sigma\,\delta\varepsilon\,\mathrm{d}v \end{aligned} \tag{7.17}$$

Thus it can be seen that the principle of virtual forces is analogous to the principle of virtual displacements discussed in Section 7.3 with the roles of forces and displacements reversed; but while the former principle represented an expression of the equilibrium conditions of a structure, the present one is a statement of its compatibility conditions or of its kinematic relations. It must again be emphasised that the virtual system—in this case the force system—is quite separate from the real (displacement) system and is in no way caused by it. Thus within the restrictions of internal and external equilibrium, any force system whatsoever can be chosen. In fact, the virtual forces do not have to be small, as long as they lie within the limitations of linear elastic theory, that forces do not cause changes in geometry. The principle holds for non-linear elasticity (as implied by fig. 7.1), but it should be remembered that for linear behaviour, work and strain energy are equal to complementary work and energy.

A general proof of the principle of virtual forces will not be included here but may be found in the references at the end of the chapter: it can, however, quite easily be shown to hold in specific cases. If, for instance, the non-uniform bar in fig. 7.7 is given an end displacement Δ_L, then some strain distribution $\varepsilon(x)$ will result within it. In terms of strain, the displacements at the end and the centre point will be

$$\Delta_L = \int_0^L \varepsilon(x)\,\mathrm{d}x$$

and

$$\Delta_{L/2} = \int_0^{L/2} \varepsilon(x)\,\mathrm{d}x$$

The virtual force system of fig. 7.7(b) is now applied to the bar. It consists of a force δP at the end and an equal and opposite force at the centre: these forces fulfil the requirement that the virtual force system should be in equili-

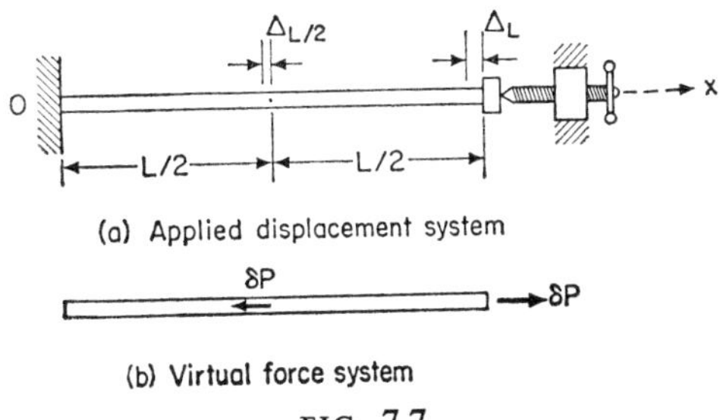

FIG. 7.7

brium. The stress at any place between the points of application of the forces will be

$$\delta\sigma = \frac{\delta P}{A}$$

where A is the local cross-sectional area of the bar. The internal complementary work is therefore

$$\delta U^* = \int_v \varepsilon(x)\,\delta\sigma\,\mathrm{d}v = \int_{L/2}^{L} \varepsilon(x)\,\frac{\delta P}{A}\,A\,\mathrm{d}x$$

and the external work is

$$\begin{aligned}\delta W^* &= \Delta_L\,\delta P - \Delta_{L/2}\,\delta P \\ &= \left(\int_0^L \varepsilon(x)\,\mathrm{d}x - \int_0^{L/2} \varepsilon(x)\,\mathrm{d}x\right)\delta P \\ &= \int_{L/2}^{L} \varepsilon(x)\,\mathrm{d}x\,\delta P\end{aligned}$$

so that

$$\delta W^* = \delta U^*$$

and the principle of virtual forces has been shown to hold. It can be shown that it also holds for a simply supported beam by the proof given on page 166 for the principle of virtual displacements, the only difference being that in this case the displacement system $u(x)$ is taken to be real while the force system $w(x)$ is virtual.

The principle of virtual forces may be used very simply to find the displacement Δ of some point of a structure if its internal displacements or strains are known. All that must be done is to apply a virtual force δP at the point at which and in the direction in which the deflection Δ is required. Statics is used to find a set of internal member forces or stresses in equilibrium with δP. For a statically determinate structure there will only be one such set of internal forces, but for an indeterminate structure more than one choice is available. If the stress at some point in a structure due to δP is

$$\delta\sigma = s\,\delta P$$

and the real strain at that point is ε, then the principle of virtual forces gives

$$\Delta\,\delta P = \int_v \varepsilon\,\delta\sigma\,\mathrm{d}v$$

$$= \int_v \varepsilon s\,\mathrm{d}v \cdot \delta P$$

or more simply

$$\Delta = \int_v \varepsilon s\,\mathrm{d}v \tag{7.18}$$

As s is the stress due to a unit force applied at the point at which Δ is required, this form of the principle is known as the *unit force method*, or sometimes, the *unit load method*.

For a pin-jointed truss, if u is the real extension of a bar and f is the bar force due to the application of a unit force where the deflection Δ is required, equation (7.18) becomes

$$\Delta = \sum fu \tag{7.19}$$

where the sum is carried out over all members. If the truss is linearly elastic and has uniform members with a section stiffness AE, and if the actual bar force in a typical member is F, then for that member

$$u = \frac{FL}{AE}$$

and (7.19) becomes

$$\Delta = \sum \frac{fFL}{AE} \tag{7.20}$$

which is the expression of the unit force method for a truss.

As a simple example, consider the truss shown in fig. 7.8 for which it is required to find the deflection at point 2 in the inclined direction shown due to the application of a vertical force of 100 (a consistent set of units is taken to apply throughout). The members of the truss are linearly elastic and the displacements under the load are small. The internal forces and hence the internal displacements in the members due to the applied load are readily calculated as shown in columns 3 and 4 of Table 7.2. A unit force is then applied at 2 in the required inclined direction and the resulting internal forces f are found. Finally, the sum of the product of f with the bar extension for each member gives the required displacement of

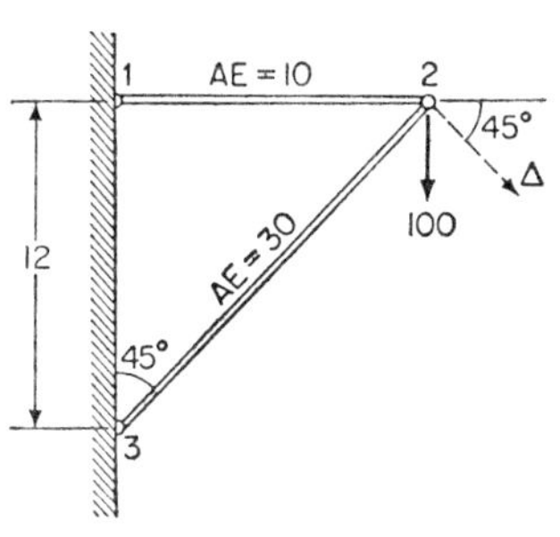

FIG. 7.8

$$\Delta = \sum \frac{fFL}{AE} = 249{\cdot}7$$

Similar expressions to (7.19) and (7.20) can be obtained for frame structures carrying their loads in bending. If the deflection Δ_i of some point i on a single beam is required, then a unit load is applied there: this will cause a bending moment m to occur in the beam. m is, of course, a variable whose value in general changes along the length of the beam. If k is the curvature of the

TABLE 7.2

1	2	3	4	5	6
Member	Member properties, L/AE	Force, F	Extension, FL/AE	Force due to unit load at 2, f	$\frac{fFL}{AE}$
1–2	1·20	+100	+120	+1·414	+169·7
2–3	0·5656	−141·4	−80	−1·0	+80·0

$$\sum \frac{fFL}{AE} = 249{\cdot}7$$

beam at some point caused by the actual load which has been applied, then equation (7.18) becomes

$$\Delta_i = \int_s mk \, ds \tag{7.21}$$

where ds is an increment of length and the integration is carried over the whole beam which, incidentally, does not have to be straight. If the beam behaves linearly, then the curvature will be

$$k = \frac{M}{EI}$$

where M is the actual bending moment in the beam, so that equation (7.21) becomes

$$\Delta_i = \int_s \frac{mM \, ds}{EI} \tag{7.22}$$

For a frame structure the expression for the deflection is the same, with the integration being carried over all the members of the structure. However, as the integrals for the different members would normally be evaluated separately, it might be helpful to re-express equation (7.22) for frame structures as

$$\Delta_i = \sum \left(\int_s \frac{mM}{EI} \, ds \right) \tag{7.23}$$

where the summation is taken over all members.

The use of the unit-force method for finding the deflection of beam and frame structures may first be illustrated by the simple example of a uniform cantilever subjected to a uniformly distributed load w (fig. 7.9(a)) for which it is required to find the end slope. Accordingly a unit moment is applied at the free end (fig. 7.9(b)) which produces a constant bending moment of $m = -1$ throughout the beam, taking sagging bending moments to be positive. The applied load will cause a bending moment of

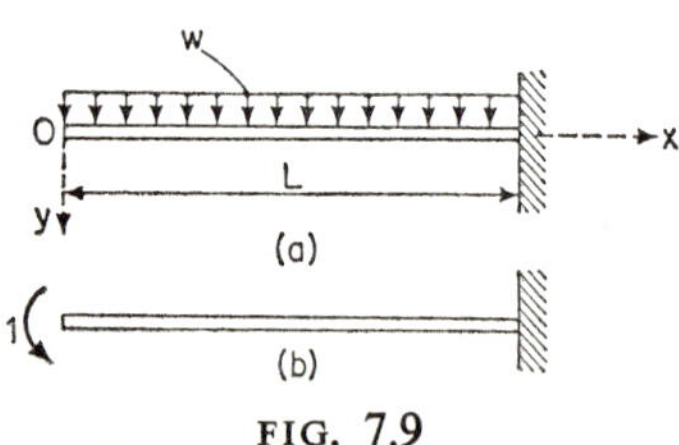

FIG. 7.9

$$M = -\frac{wx^2}{2}$$

to occur, and hence by equation (7.22) the required slope will be

$$\theta_0 = \int_0^L \frac{mM \, dx}{EI} = \int_0^L 1 . \frac{wx^2}{2EI} \, dx = \frac{wL^3}{6EI}$$

If the vertical deflection of the end is required, a vertical unit force is applied at 0. This will cause a virtual moment within the beam of

$$m = -x$$

(if the unit force is applied downwards), and the deflection will be

$$\Delta_0 = \int_0^L \frac{mM}{EI}\,dx = \int_0^L x\frac{wx^2}{2EI}\,dx = \frac{wL^4}{8EI}$$

So far, only statically determinate structures have been considered. However, the principle of virtual forces may be used to express the overall kinematic relationships of a structure, and for a statically indeterminate structure these must include its compatibility conditions. Hence the principle may be used for the analysis of statically indeterminate structures by the force method. For a two-span beam its use in this way may be illustrated as follows.

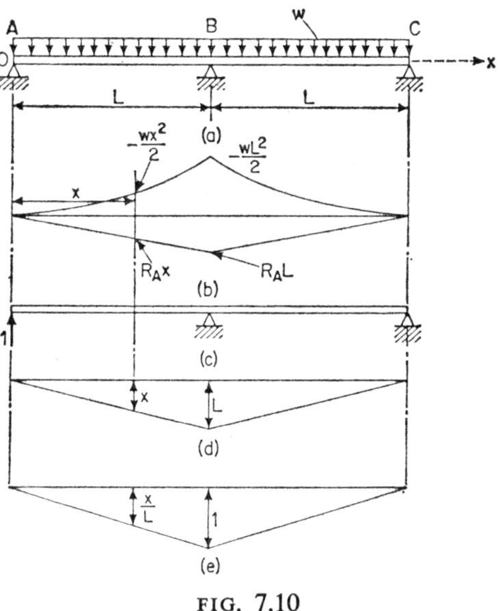

FIG. 7.10

It is required to analyse the beam shown in fig. 7.10(a): the beam is uniform and is subjected to a uniformly distributed load w. It is necessary first to express the bending moment M at any point in terms of the applied load and the redundant forces—in this case there is only one redundant force as the structure is one times statically indeterminate. Choosing the redundant as the reaction R_A at A, then the bending moments due to w and to R_A may be drawn separately as in fig. 7.10(b). For the left-hand span, it can be seen that the total bending moment at any point is

$$M = R_A x - \frac{wx^2}{2}$$

The compatibility equation required for the solution of the problem may be obtained by calculating the deflection of A and equating it to zero. This deflection may be found by applying a unit force to the beam at A (fig. 7.10(c)) which produces the virtual bending moment distribution shown in fig. 7.10(d). For the left-hand span this will have a value

$$m = x$$

Then by equation (7.22),

$$\Delta_A = \int_s \frac{mM}{EI}\,\mathrm{d}s$$

$$= 2\int_0^L \frac{x(R_A x - wx^2/2)\,\mathrm{d}x}{EI}$$

$$= \frac{2}{EI}\left(\frac{R_A L^3}{3} - \frac{wL^4}{8}\right) = 0$$

Hence the result is

$$R_A = \tfrac{3}{8}wL$$

Note that the compatibility relation could be thought of in a somewhat different way by returning to equation (7.16). It can be seen that the external complementary virtual work δW^* done by the virtual force system of fig. 7.10(d) will be zero as point A has no displacement, so that equation (7.16) reduces to

$$\delta U^* = 0$$

which, in this case, can be written

$$\int_s \frac{mM}{EI}\,\mathrm{d}s = 0$$

as before.

In fig. 7.10(c), the unit force was applied at A, in the same direction as the unknown reaction R_A. However, the position of the unit force is not restricted to that of the redundant: a unit force could be applied anywhere as long as it could be used to describe a compatibility condition; and by the principle of virtual forces, any force distribution satisfying equilibrium may be chosen. For example, the unit force could be chosen to be the bending moment at point B, implying a compatibility condition that no relative rotation between the two halves of the beam occurs. The corresponding virtual bending moment diagram is shown in fig. 7.10(e), and as no external work is done the compatibility equation becomes

$$0 = \int_0^{2L} \frac{mM}{EI}\,\mathrm{d}x = \frac{2}{EI}\int_0^L \frac{x}{L}\left(R_A x - \frac{wx^2}{2}\right)\mathrm{d}x$$

$$= \frac{2}{EI}\left(\frac{R_A L^2}{3} - \frac{wL^3}{8}\right)$$

so that

$$R_A = \tfrac{3}{8}wL$$

as before. It will be noticed, nevertheless, that despite the fact that the unit force was applied in different places, the bending moment diagrams of fig. 7.10(d and e) have the same shape even though their magnitudes are different. This similarity of shape occurs because the structure is only one times statically indeterminate, which means that there can be only one internal self-equilibrating force (bending moment) distribution.

The three-times indeterminate frame of fig. 7.11 may be analysed in much the same way, but because of the large number of terms involved it is essential to carry out such an analysis in tabular form. First, the reactions R_1, R_2 and R_3 (fig. 7.11(b)) are chosen as the redundant force system. Unit forces will

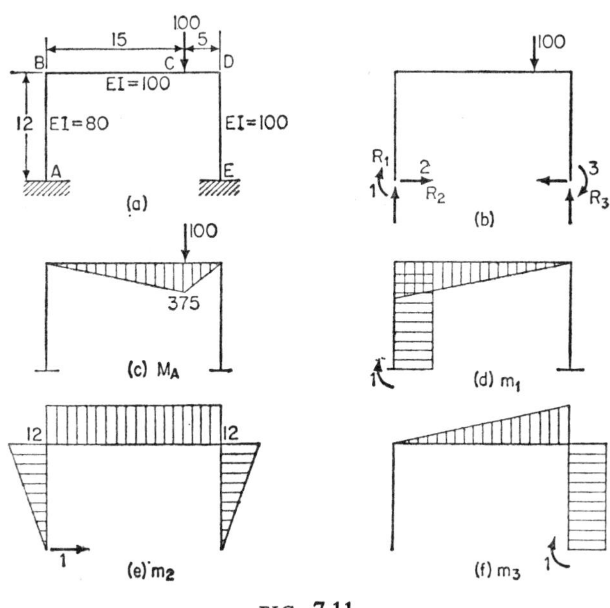

FIG. 7.11

also be applied to the statically determinate structure at point-directions 1, 2 and 3 to find the displacements there: these will of course be zero, so giving the required compatibility equations. Fig. 7.11(c) shows the bending moment distribution M_A when the load of 100 has been applied to the statically determinate structure, and fig. 7.11(d, e and f) show the bending moments m_1, m_2 and m_3 due to unit forces applied at points 1, 2 and 3. These values are given in Table 7.3. Note that the sign convention used is that bending moments causing the frame to be concave inwards are taken as positive: it is extremely important in this type of problem to establish a sign convention and then to follow it rigorously. The bending moments due to the redundant forces will of course be given by the bending moment due to a unit force multiplied by the appropriate redundant: for instance, the bending moment due to R_2 is

$$M_2 = R_2 m_2$$

The product Mm and the complementary work $\int_s (Mm/EI)\,\mathrm{d}s$ are now calculated in Table 7.4 for each member, terms being produced by three unit forces acting through the displacements due to the applied loads and due to the three redundant forces in turn. As M_A is not a continuous function over the whole of BD, the product mM_A has to be integrated separately for BC and CD.

TABLE 7.3

<table>
<tr><th>1</th><th>2</th><th>3</th><th>4</th><th>5</th><th>6</th></tr>
<tr><th>Member</th><th>EI</th><th>M_A</th><th>m_1</th><th>m_2</th><th>m_3</th></tr>
<tr><td>AB</td><td>80</td><td>0</td><td>1</td><td>$-y$</td><td>0</td></tr>
<tr><td>BC</td><td>100</td><td>$\frac{375x}{15} = 25x$</td><td rowspan="2">$\frac{20-x}{20}$</td><td rowspan="2">-12</td><td rowspan="2">$\frac{-x}{20}$</td></tr>
<tr><td>CD</td><td>100</td><td>$\frac{365(20-x)}{5} = 75(20-x)$</td></tr>
<tr><td>DE</td><td>100</td><td>0</td><td>0</td><td>$-y$</td><td>-1</td></tr>
</table>

The compatibility equations may now be written. The displacement at 1, which is zero for compatibility, is given by

$$\Sigma\left(\int\frac{m_1 M_A\,\mathrm{d}s}{EI}\right) + \Sigma\left(\int\frac{m_1 M_1\,\mathrm{d}s}{EI}\right) + \Sigma\left(\int\frac{m_1 M_2\,\mathrm{d}s}{EI}\right) + \Sigma\left(\int\frac{m_1 M_3\,\mathrm{d}s}{EI}\right) = 0$$

Summing the appropriate terms in Table 7.4, this becomes

$$15{\cdot}64 + 2{\cdot}167R_1 - 2{\cdot}1R_2 - 0{\cdot}0333R_3 = 0$$

With the applied load terms transferred to the right-hand side, this together with the other two compatibility equations can be written in matrix form as

$$\begin{bmatrix} 0{\cdot}2167 & -2{\cdot}1 & -0{\cdot}0333 \\ -2{\cdot}1 & 41{\cdot}76 & 1{\cdot}92 \\ -0{\cdot}0333 & 1{\cdot}92 & 0{\cdot}1867 \end{bmatrix} \begin{Bmatrix} R_1 \\ R_2 \\ R_3 \end{Bmatrix} = \begin{Bmatrix} -15{\cdot}64 \\ +450 \\ +21{\cdot}87 \end{Bmatrix}$$

TABLE 7.4

1	2	3	4	5	6	7	8	9	10
Unit load at	Member	Applied load		R_1		R_2		R_3	
		mM_A	$\int \frac{mM_A\,ds}{EI}$	mM_1	$\int \frac{mM_1\,ds}{EI}$	mM_2	$\int \frac{mM_2\,ds}{EI}$	mM_3	$\int \frac{mM_3\,ds}{EI}$
1	AB	0	0	R_1	$0{\cdot}15R_1$	$-yR_2$	$-0{\cdot}9R_2$	0	0
	BC	$\frac{25x}{20}(20-x)$	14·8	$\left(\frac{20-x}{20}\right)^2 R_1$	$0{\cdot}0667R_1$	$-\frac{12(20-x)}{20}R_2$	$-1{\cdot}2R_2$	$-\frac{(20-x)x}{400}R_3$	$-0{\cdot}0333R_3$
	CD	$\frac{75}{20}(20-x)^2$	1·56						
	DE	0	0	0	0	0	0	0	0
2	AB	0	0	$-yR_1$	$-0{\cdot}9R_1$	y^2R_2	$7{\cdot}2R_2$	0	0
	BC	$-300x$	$-337{\cdot}5$	$-\frac{12}{20}(20-x)R_1$	$-1{\cdot}2R_1$	$144R_2$	$28{\cdot}8R_2$	$\frac{12}{20}xR_3$	$1{\cdot}2R_3$
	CD	$-900(20-x)$	$-112{\cdot}5$						
	DE	0	0	0	0	y^2R_2	$5{\cdot}66R_2$	yR_3	$0{\cdot}72R_3$
3	AB	0	0	0	0	0	0	0	0
	BC	$-\frac{25}{20}x_2$	$-14{\cdot}05$	$-\frac{x(20-x)}{400}R_1$	$-0{\cdot}0333R_1$	$\frac{12}{20}xR_2$	$1{\cdot}2R_2$	$\frac{x^2}{400}R_3$	$0{\cdot}0667R_3$
	CD	$-\frac{75x}{20}(20-x)$	$-7{\cdot}82$						
	DE	0	0	0	0	yR_2	$0{\cdot}72R_2$	R_3	$0{\cdot}12R_3$

The solution of these equations gives

$$R_1 = 88{\cdot}2$$
$$R_2 = 17{\cdot}27$$
$$R_3 = -44{\cdot}7$$

which agrees with previous results for the same structure given in Chapters 5 and 6.

Internal forces can also arise in a statically indeterminate structure due to self-strain, due, perhaps, to an initial lack of fit when assembling the structure, to thermally induced strains and so on. Such problems can be dealt with quite well using the unit-force method. The approach used will be illustrated by considering the pin-jointed truss of fig. 7.12, though the same basic ideas can be applied equally well to general frame structures.

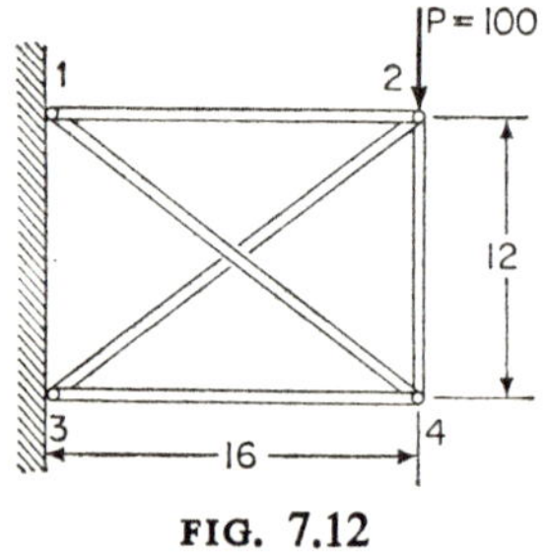

FIG. 7.12

The two diagonal members of the truss have a section stiffness of $AE = 300$ units, while for the other three members $AE = 200$. During assembly, member 1–4 was inserted last and proved to be 5 units too short. It is required to find the forces in the members and the vertical deflection of point 4 when a force of 100 is applied at 2 as shown.

As usual, the first step is to select a redundant force system and so define a statically determinate structure. In this case, the force F_{14} in member 1–4 is chosen as the unknown. The forces in all

TABLE 7.5

1	2	3	4		5	
Member	Length	EA	Forces F due to		Extensions FL/AE due to	
			P	F_{14}	$P = 100$	F_{14}
1–2	16	200	$1{\cdot}33P$	$-0{\cdot}8F_{14}$	$0{\cdot}1064P$	$-0{\cdot}064F_{14}$
2–4	12	200	0	$-0{\cdot}6F_{14}$	0	$-0{\cdot}036F_{14}$
3–4	16	200	0	$-0{\cdot}8F_{14}$	0	$-0{\cdot}064F_{14}$
1–4	20	300	0	$+F_{14}$	0	$+0{\cdot}0667F_{14}$
2–3	20	300	$-1{\cdot}67P$	$+F_{14}$	$-0{\cdot}1111P$	$+0{\cdot}0667F_{14}$
					TOTALS →	

members are now calculated by statics and their values entered in column 4 of Table 7.5. Those due to the applied load are expressed algebraically in terms of P for reasons which will become apparent later. In column 5 the member extensions due to these forces are calculated. The compatibility condition must now be established from which the unknown force F_{14} can be found, which may be done as follows. If member 1–4 were cut, then if no loads were applied to the structure the member would have a gap in it of 5 units between the cut ends (fig. 7.13(a)). We know, however, that if the load P at 2 were applied to the structure and if a force (which can be thought of as an *external* force) of F_{14} were applied to each side of the cut, then the two ends would move together and exactly meet for compatibility (fig. 7.13(b)). If the left- and right-hand sides of the cut moved by amounts Δ_L and Δ_R, then these quantities could be found by applying a unit force first to the right- and then to the left-hand sides of the cut. The total internal work caused by each would give the appropriate displacement. However, the

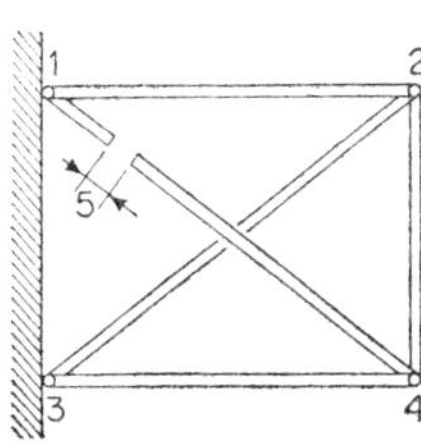

(a) No forces on structure
Gap of 5 in 1–4

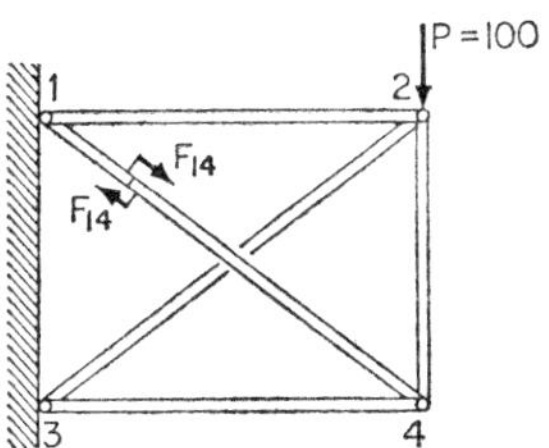

(b) Forces P and F_{14} on structure
Gap is zero for compatibility

FIG. 7.13

6	7	8	9
Forces f^1 due to unit force in 1–4	Work $\delta U_1^* = f^1(FL/AE)$	Forces f^2 due to unit force at 4	Work $\delta U_2^* = f^2(FL/AE)$
$-0{\cdot}8$	$-0{\cdot}0851P + 0{\cdot}0512F_{14}$	$1{\cdot}33$	$+0{\cdot}142P - 0{\cdot}0852F_{14}$
$-0{\cdot}6$	$+ 0{\cdot}0216F_{14}$	$1{\cdot}0$	$- 0{\cdot}0360F_{14}$
$-0{\cdot}8$	$+ 0{\cdot}0512F_{14}$	0	0
1	$+ 0{\cdot}0667F_{14}$	0	0
1	$-0{\cdot}1111P + 0{\cdot}0667F_{14}$	$-1{\cdot}67$	$+0{\cdot}1853P - 0{\cdot}1111F_{14}$
$\delta U_1^* = -0{\cdot}1962P + 0{\cdot}2574F_{14}$		$\delta U_2^* = +0{\cdot}3273P - 0{\cdot}2323F_{14}$	

individual displacements are not in fact required, for the compatibility condition for the member (and the structure) is that

$$\Delta_L + \Delta_R = 5$$

It is therefore the sum of the two displacements that is needed, a quantity which is obtained if unit forces are applied to both sides of the cut simultaneously—or if, in other words, a unit *internal* force is assigned to member 1–4. The internal forces which result in this and the other members are found in column 6 of Table 7.5, the internal complementary work in each member is calculated in column 7 and the sum is

$$\delta U_1^* = -0{\cdot}1962P + 0{\cdot}2574F_{14}$$

The external work is

$$\begin{aligned}\delta W_1^* &= 1.\Delta_L + 1.\Delta_R \\ &= 1.(\Delta_L + \Delta_R) = 5\end{aligned}$$

Hence by the principle of virtual forces,

$$\delta W_1^* = \delta U_1^*$$

and so

$$-0{\cdot}1962P + 0{\cdot}2574F_{14} = 5 \tag{7.24}$$

which is the required compatibility equation. Hence we can say that if the initial lack of fit in a member is t, then if a unit force is assigned to that member, the resulting complementary work done within the structure is equal to the lack of fit, or

$$\delta U^* = t \tag{7.25}$$

In the present case, putting $P = 100$ in equation (7.24) gives the result

$$F_{14} = 95{\cdot}8$$

which is the force in the bar due to P and to the initial lack of fit. If, in a similar problem, it is not immediately clear what sign to give to the lack of fit in equation (7.25) (should the sign be positive if the bar is too long, or if it is too short?) then the applied load should be set to zero: the sign should then become obvious. In this case, for instance, if P is zero, we know that the force in 1–4 must be tensile as the bar was originally too short.

The next part of the problem is the determination of the deflection of point 4 due to the application of P to point 2. This may be done by applying a unit vertical force at 4 in the required direction (column 8, Table 7.5) and calculating the complementary internal work that results (column 9): this quantity will, of course, give the required deflection, or

$$\Delta_4 = \delta U_2^* = +0{\cdot}3273P - 0{\cdot}2323F_{14} \tag{7.26}$$

Care must be taken at this point, however, for what is required is the deflection of 4 due only to the application of P, without including any effect due to the lack of fit of member 1–4. But the value of F_{14} calculated above includes the self-strain effect, so if it were used in equation (7.26) the resulting deflection would be that occurring from the initial position held by point 4 before member 1–4 had been inserted. Consequently, the component of F_{14} due only to P and not to self-strain is required for equation (7.26): this may be found by setting the lack of fit equal to zero in (7.24) which gives

$$F_{14}^P = \frac{0{\cdot}1962 \times 100}{0{\cdot}2574} = 76{\cdot}3$$

Hence

$$\Delta_4 = 0{\cdot}3273 \times 100 - 0{\cdot}2323 \times 76{\cdot}3 = 15{\cdot}0$$

The deflection of 4 due to the lack of fit of 1–4 could be found quite simply if required by putting $P = 0$ in both equations (7.24) and (7.26). In such a case,

$$\Delta_4 = -0{\cdot}2323 \times \frac{5}{0{\cdot}2574} = -4{\cdot}52$$

7.6 Castigliano's Theorem, Part II

Fig. 7.14 represents an elastic structure (which is not necessarily linear) subjected to a series of forces $P_1, P_2, \ldots$ These cause displacements $\Delta_1, \Delta_2 \ldots$ to occur, and the complementary energy U^* of the structure due to the forces can be calculated. If now a small change δP_i is made in force P_i, with all the other forces remaining unaltered, then if δP_i is small enough not to have affected the magnitude of the displacements, the complementary work done on the structure will be

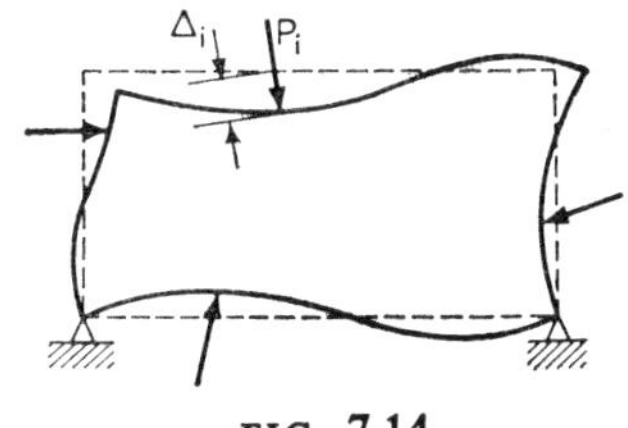

FIG. 7.14

$$\delta W^* = \Delta_i \, \delta P_i$$

which may be written, from the principle of virtual forces, as

$$\delta U^* = \Delta_i \, \delta P_i$$

In the limit, as $\delta P_i \to 0$, this becomes

$$\Delta_i = \frac{\partial U^*}{\partial P_i} \tag{7.27}$$

This may be expressed in the form:

> If an elastic structure is acted upon by a series of forces $P_1, P_2 \ldots$, then the displacement of point i is equal to the partial derivative of the complementary energy of the structure with regard to the force at i.

The complementary nature of equations (7.14) and (7.27) should be noted. For a linear elastic structure the strain energy and complementary energy are equal so that equation (7.27) becomes

$$\Delta_i = \frac{\partial U}{\partial P_i} \tag{7.28}$$

This is the form first derived by Castigliano, known as *Castigliano's theorem, Part II*, and as all further discussion in this chapter will be confined to linearly elastic structures, it is this form rather than the stricter complementary energy form which will be used from now on.

If a displacement is required at some point at which no external force occurs, an additional 'dummy' force must be applied there. The displacement can then be calculated using equation (7.28), and the force equated to zero.

Although equation (7.28) appears to be formulated in terms of strain energy, the strain energy of a structure need never actually be calculated, for only its derivative is required. For example, the strain energy in some bar j–k of a pin-jointed truss is $\frac{1}{2}(F_{jk}^2 L_{jk}/AE)$ but as we actually only need

$$\frac{\partial}{\partial P_i}\left(\frac{1}{2}\cdot\frac{F_{jk}^2 L_{jk}}{AE}\right) = \frac{F_{jk}L_{jk}}{AE}\cdot\frac{\partial F_{jk}}{\partial P_i}$$

we might as well calculate this quantity directly. Note that $F_{jk}L_{jk}/AE$ is the extension of member j–k due to all the external forces and internal self-strain of the truss—a term essential to the unit-force method of analysis discussed in the previous section. The member force F_{jk} is itself a linear function of all the applied forces P_i and redundant forces R_i and has the form

$$F_{jk} = a_1P_1 + a_2P_2 + \cdots + a_iP_i + \cdots + b_1R_1 + b_2R_2 + \cdots$$

where a_m, b_m are constants determined by statics. Now

$$\frac{\partial F_{jk}}{\partial P_i} = a_i$$

which is simply the force in j–k due to a unit force applied at i. Thus Castigliano's theorem, Part II, although appearing to be quite different from the principle of virtual forces, also reduces to the unit force method, and the calculations involved in an analysis are exactly the same as those which would be used if the principle of virtual forces were applied.

For example, the calculations involved for the truss of fig. 7.12 would be the same as those given in Table 7.5. The forces in column 8 would be those due to a (unit) dummy force applied at point 4 (or, calling the dummy load P_4, they would be equal to $\partial F_{jk}/\partial P_4$). The compatibility equation and the method of dealing with the lack of fit would also be exactly the same as in the previous example, for the unit-load method is the same no matter whether it is based directly on the principle of virtual forces or on Castigliano's theorem, Part II.

For a beam, the strain energy due to bending is

$$U = \int_0^L \frac{M^2}{2EI}\,\mathrm{d}s$$

as at any section the bending moment is M and the curvature is M/EI. Then

$$\frac{\partial U}{\partial P_i} = \Delta_i = \int_0^L \frac{M}{EI}\cdot\frac{\partial M}{\partial P_i}\,\mathrm{d}s \tag{7.29}$$

which may be evaluated directly. But just as for the truss, the bending moment in a beam is a linear function of the applied forces and redundants, so that in the same way, the term $\partial M/\partial P_i$ is equal to the bending moment m due to a unit force applied at i. Hence once again Castigliano's theorem, Part II, reduces to the unit force formulation, which is, in the case of a frame structure,

$$\Delta_i = \sum\left(\int_0^L \frac{mM}{EI}\,\mathrm{d}s\right)$$

where the sum is carried out over all members.

Many other energy theorems exist in structural analysis which are dealt with in references 1 and 2 below: they are well worth pursuing as background reading in the subject of work and energy methods. Perhaps the most interesting are the Rayleigh-Ritz and Galerkin methods of using strain energy to obtain approximate solutions: to some extent, as has been shown especially in Chapter 5, even 'exact' methods are necessarily approximate so that a deliberately approximate approach is often realistic, especially if the bounds of the solution (whether the obtained result is above or below the true result for the structure) are known. It is in such directions that the greatest value of energy methods lies.

References

1. Argyris, J. H., and Kelsey, S., *Energy Theorems and Structural Analysis*, Butterworth, London, 1960.
2. Neal, B. G., *Structural Theorems and their Applications*, Pergamon, Oxford, 1964.

PROBLEMS

7.1 Find the flexibility matrix for the two directions shown at point E of the plane, pin-jointed truss given in fig. 7.15. All members have the same cross-sectional area of 0·2, and $E = 3 \times 10^9$.

Answer: $\begin{bmatrix} 69{\cdot}0 & 23{\cdot}3 \\ 23{\cdot}3 & 11{\cdot}7 \end{bmatrix} \times 10^{-7}$

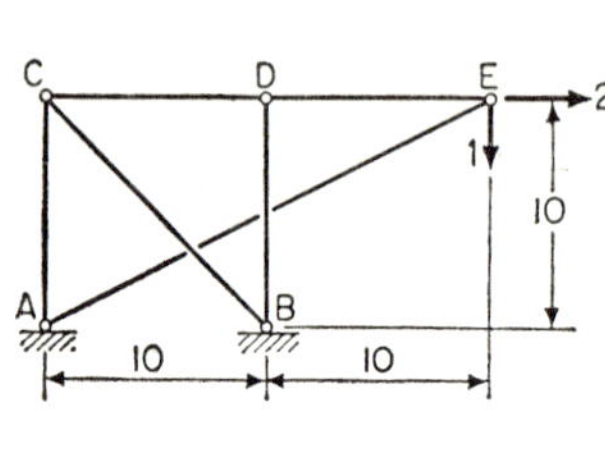

FIG. 7.15

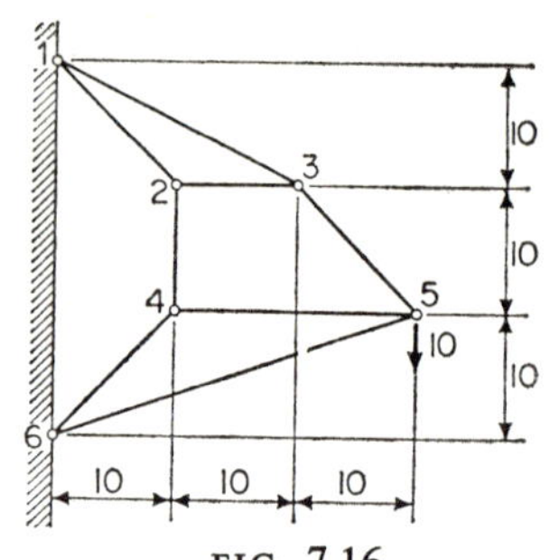

FIG. 7.16

7.2 All members of the statically determinate pin-jointed truss shown in fig. 7.16 have a cross-sectional area of 0·1 and a Young's modulus of of 3×10^9. It is decided to modify the design by increasing the cross-sectional area of member 3–5 to 0·2. Find the change caused by this modification in the vertical deflection of point 5 due to a vertical load of 10,000 applied at that point.

Answer: 0·00566

7.3 The section stiffness *EI* at one end of a beam is double its values at the other. $1/EI$ varies linearly along the length L of the beam. Use an energy method to find the bending moments that must be applied to each end in turn to cause unit rotation there when the remote end is fully fixed.

7.4 For the frame of fig. 7.17, use an energy method to determine
(a) the horizontal deflection of B
(b) the horizontal deflection of D
(c) the rotation at D.
Check your results by the method of superposition.

7.5 The plane frame shown in fig. 7.18 is pin-jointed at the lettered points. All the members have the same cross-section. Determine the force in each member due to the application of the load of 7·0 at *E*.

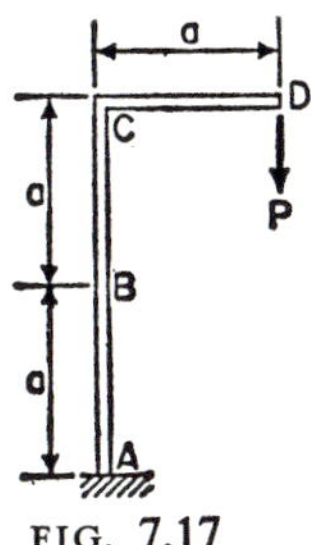

FIG. 7.17

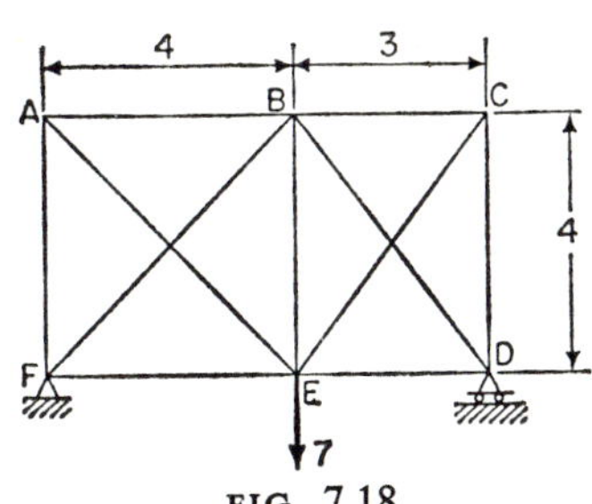

FIG. 7.18

7.6 Strain gauges are mounted on member 2–4 of the pin-jointed truss shown in fig. 7.19. When a load of 9000 is applied at joint 5, as shown, the gauges show a compression force of 8000 in member 2–4. During the initial erection of the truss (that is, before application of the load of 9000), find whether member 2–4 was too long or too short, and by how much. All members have cross-sectional areas of 0·2. Young's modulus is 3×10^9.

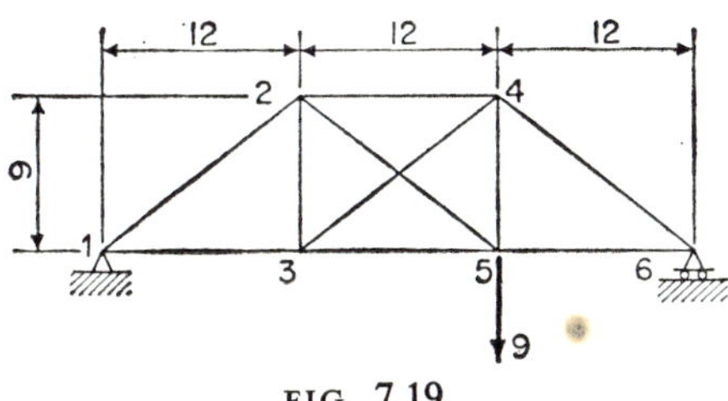

FIG. 7.19

Answer: 0·00203

7.7 The pin-ended truss shown in fig. 7.20 consists of three light alloy members AB, BD and DC, for which (in the usual notation) $E = 10^9$ and $A = 0{\cdot}12$. The two cross bracing members AD and BC are steel wires for which $E = 3 \times 10^9$ and $A = 0{\cdot}02$. With no external loads applied to the truss, the steel wires are tensioned to a force of 2000. A vertical force P is then applied at D. Draw a load/deflection curve of P against the vertical deflection of D as P is increased from zero to 5000, assuming that the wires can take no load in compression.

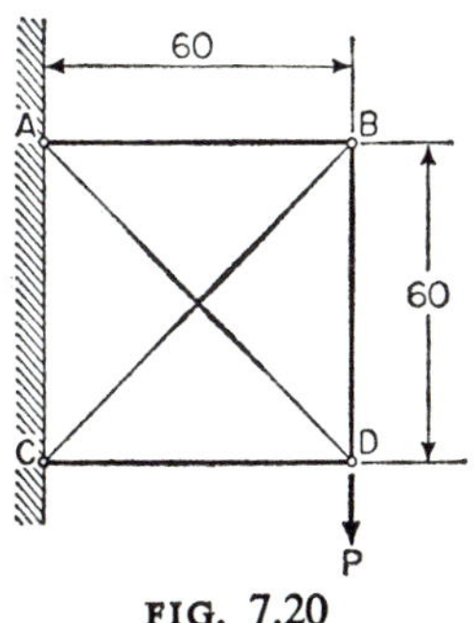

FIG. 7.20

Answer: Final deflection = 0·0119

7.8 Use an energy method to find the moments in terms of P and L at points 1 and 6 of the symmetrically loaded frame shown in fig. 7.21. All members are uniform and have the same sectional properties.

Answer: $M_1 = 0{\cdot}298PL$

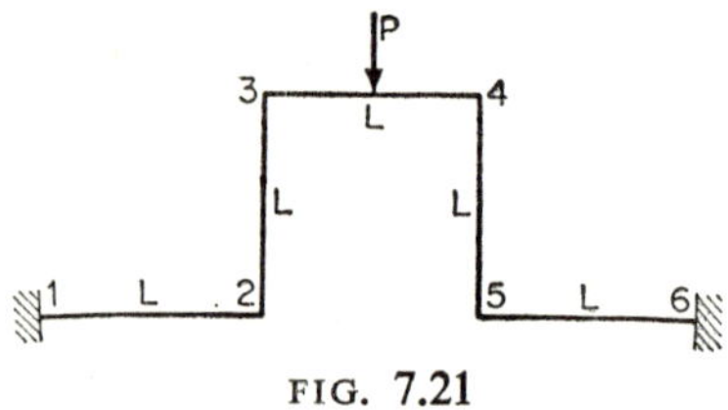

FIG. 7.21

7.9 Do problem 6.11, using an energy method.

8

Matrix Analysis

8.1 Introduction

The general discussion on structural analysis given in Chapter 2 explained that wherever possible, structures should be broken down into a number of discrete elements for the purpose of analysis. The analysis would then proceed by first evaluating the stiffness or flexibility characteristics of the individual elements, followed by assembling the elements together and using conditions of compatibility or equilibrium for the final solution by the force or displacement method. Matrix analysis follows this procedure precisely: indeed, in a sense matrix analysis has already been used earlier in this book in an informal way, whenever compatibility or stiffness equations have been expressed in matrix form. This chapter deals with the process more formally, however, and uses matrix notation to present in a logical form the assembly of element stiffness and flexibility matrices into stiffness and flexibility matrices for the whole structure, and to show how the force and displacement methods may be expressed in a general way applicable to all structures.

For reasons of space, the chapter has been confined to dealing only with one-dimensional linear (bar or beam) structural elements each connecting two nodes only, so that only frame or truss structures are discussed. However, continuum problems can be dealt with in exactly the same way using two- or three-dimensional structural elements: these are generally called *finite elements*. A plate, a diaphragm or a shell, for instance, could be divided into triangular or quadrilateral finite elements; and although the calculation of the individual element stiffness characteristics is considerably more complicated than for the one-dimensional elements of this chapter, the technique of assembling the elements to obtain the behaviour of the complete structure is virtually the same.

One point should be realised at this stage: the formal statements of matrix analysis are seldom if ever followed in practice. They are too cumbersome for use by hand and too inefficient in time and storage for use as they stand by computer. They are, however, logical and rigorous, and once understood they can be shortened and used as the basis of any type of elastic structural analysis though their use is, of course, primarily in the formulation of computer solutions.

8.2 Definitions and Sign Conventions

In general, two types of rectangular Cartesian coordinate systems will be used: *member* or *local* coordinates (x, y, z) oriented along a member, and *system* coordinates (X, Y, Z) which are fixed with regard to a specified origin and axis system for the entire structure. Quantities (forces, displacements, flexibility and stiffness coefficients and matrices) which are specified in member coordinates will be denoted by lower case type, while quantities in system coordinates will be written in upper case characters.

Linear displacements are denoted by u, v, w in the x, y and z directions, and rotational displacements by θ, φ and ψ as shown in fig. 8.1(a) (lower case

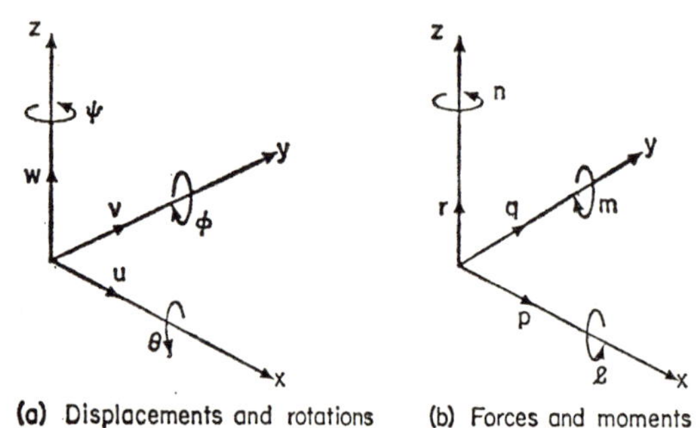

FIG. 8.1 Three-dimensional force and displacement notations

letters are used in this instance as if member coordinates were being used—in system coordinates, of course, the symbols would be U, V, W, Θ, Φ and Ψ). The notation for forces (p, q, r) and moments (l, m, n) is given in fig. 8.1(b). It will be seen that a right-hand convention is used: this will be adhered to throughout. Fig. 8.2 gives two alternate two-dimensional systems of displacement and force compatible with the three-dimensional systems of the previous figure. Forces acting on the ends of members are differentiated from external forces applied to nodes by the number of subscripts used: this is illustrated in fig. 8.3 in which, for example, p_{ji} and P_{ji} are forces acting on the end j of member i–j in the x and X directions respectively, while P_j is an external load applied in the X-direction at node j. Member-end and nodal displacements follow much the same convention except that displacements with a single subscript refer to *absolute* nodal displacements whereas a double subscript indicates a *relative* displacement of one end of a member relative to the other. Thus W_j is the displacement of node j in the Z-direction while v_{ij}, V_{ij} refer to the relative displacements of i with regard to j for member ij in the y or Y directions. Relative displacements will also be referred to as *deformations*.

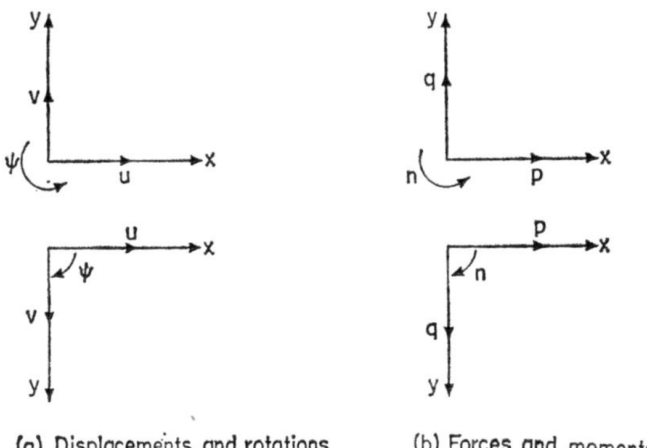

FIG. 8.2 Alternative two-dimensional force and displacement systems

The notation used earlier in the book will be retained, that matrices of any kind will be written in bold-face type thus: **K**; and that row vectors, rectangular matrices and column vectors may be denoted by the use of the brackets ⌊ ⌋, [ ] and { } respectively. Force and displacement vectors will in

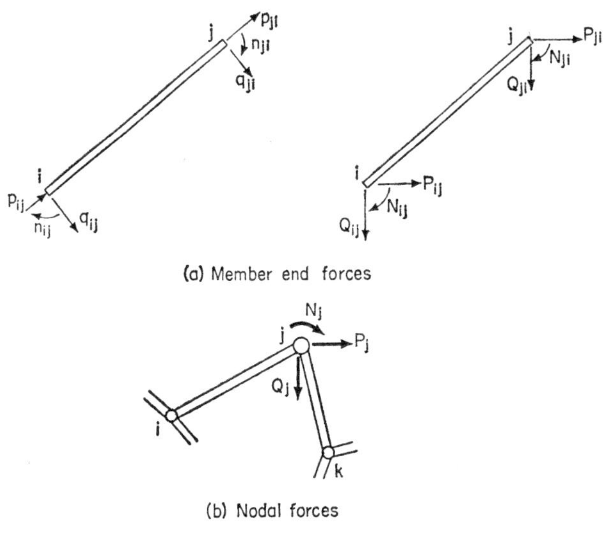

FIG. 8.3

general be described by the terms **P** and **Δ**: for one node i loaded and displaced in all six possible directions, these vectors will be

$$\mathbf{P}_i = \begin{Bmatrix} P_i \\ Q_i \\ R_i \\ L_i \\ M_i \\ N_i \end{Bmatrix} \quad \text{and} \quad \mathbf{\Delta}_i = \begin{Bmatrix} U_i \\ V_i \\ W_i \\ \Theta_i \\ \Phi_i \\ \Psi_i \end{Bmatrix} \tag{8.1}$$

The order in which the coefficients of these vectors are written should be noted carefully: this order will always be followed even when individual terms are missing and the matrices are no longer 6 × 1 vectors. The order in which forces and displacements are written will also, of course, dictate the order of the coefficients in the corresponding flexibility and stiffness matrices.

8.3 Element Flexibility and Stiffness Matrices

For a prismatic beam of symmetrical section fixed at one end as shown in fig. 8.4(a), for which forces and moments are applied at the centroid of the section and for which the y and z axes are aligned with the principal axes of the section, the flexibility and stiffness matrices will be, following the order of coefficients in equation (8.1),

$$\mathbf{f} = \left[\begin{array}{ccc|ccc} \dfrac{L}{AE} & 0 & 0 & 0 & 0 & 0 \\ 0 & \dfrac{L^3}{3EI_{zz}} & 0 & 0 & 0 & \dfrac{L^2}{2EI_{zz}} \\ 0 & 0 & \dfrac{L^3}{3EI_{yy}} & 0 & -\dfrac{L^2}{2EI_{yy}} & 0 \\ \hline 0 & 0 & 0 & \dfrac{L}{GJ} & 0 & 0 \\ 0 & 0 & -\dfrac{L^2}{2EI_{yy}} & 0 & \dfrac{L}{EI_{yy}} & 0 \\ 0 & \dfrac{L^2}{2EI_{zz}} & 0 & 0 & 0 & \dfrac{L}{EI_{zz}} \end{array}\right] \tag{8.2}$$

$$\mathbf{k} = \left[\begin{array}{ccc|ccc} \dfrac{AE}{L} & 0 & 0 & 0 & 0 & 0 \\ 0 & \dfrac{12EI_{zz}}{L^3} & 0 & 0 & 0 & -\dfrac{6EI_{zz}}{L^2} \\ 0 & 0 & \dfrac{12EI_{yy}}{L^3} & 0 & \dfrac{6EI_{yy}}{L^2} & 0 \\ \hline 0 & 0 & 0 & \dfrac{GJ}{L} & 0 & 0 \\ 0 & 0 & \dfrac{6EI_{yy}}{L^2} & 0 & \dfrac{4EI_{yy}}{L} & 0 \\ 0 & -\dfrac{6EI_{zz}}{L^2} & 0 & 0 & 0 & \dfrac{4EI_{zz}}{L} \end{array}\right] \tag{8.3}$$

The various material and geometrical properties of the beam are listed above it in the figure, and I_{yy} and I_{zz} are the second moments of area of the section about the y and z axes. It should be remembered, however, that although the flexibility and stiffness matrices of equations (8.2) and (8.3) will apply to all beams dealt with in this chapter, for the general case of a cantilever which is

FIG. 8.4

neither straight nor uniform and which is loaded in an arbitrary direction, the stiffness and flexibility matrices will consist entirely of non-zero elements.

If the two-dimensional behaviour alone of the beam of fig. 8.4(a) is required, then only the first, second and last rows and columns of the three-dimensional flexibility and stiffness matrices need be retained, leaving for the beam of fig. 8.4(b),

$$\mathbf{f} = \begin{bmatrix} \dfrac{L}{AE} & 0 & 0 \\ 0 & \dfrac{L^3}{3EI} & \dfrac{L^2}{2EI} \\ 0 & \dfrac{L^2}{2EI} & \dfrac{L}{EI} \end{bmatrix}; \quad \mathbf{k} = \begin{bmatrix} \dfrac{AE}{L} & 0 & 0 \\ 0 & \dfrac{12EI}{L^3} & -\dfrac{6EI}{L^2} \\ 0 & -\dfrac{6EI}{L^2} & \dfrac{4EI}{L} \end{bmatrix} \tag{8.4}$$

8.4 Translation Matrices

Fig. 8.5 shows two elements (m) and (n) connecting points i, j and k. The elements are shown connected in fig. 8.5(a), while in (b) they are shown separated with connecting forces in between. It is required to find the force

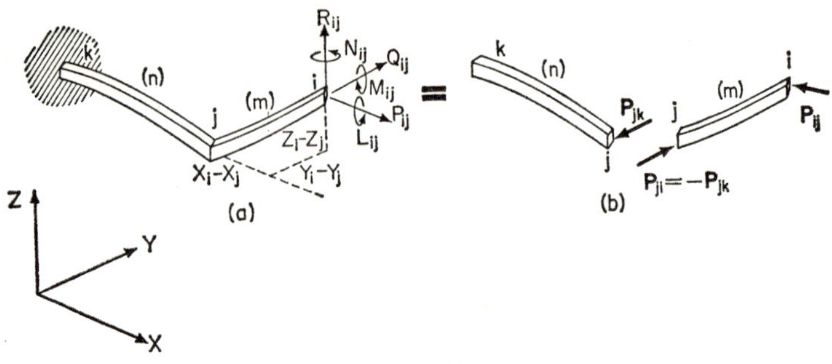

FIG. 8.5

system $\mathbf{P}_{jk}$ acting on member (n) at j due to a force system $\mathbf{P}_{ij}$ applied to (m) at i: in fig. 8.5(a), $\mathbf{P}_{ij}$ is shown resolved into its six separate components. The relationship between $\mathbf{P}_{ij}$ and $\mathbf{P}_{jk}$ could be written as

$$\mathbf{P}_{jk} = \mathbf{H}_{ji}\mathbf{P}_{ij} \tag{8.5}$$

where by statics it can be seen that

$$\mathbf{H}_{ji} = \left[\begin{array}{ccc:ccc}
1 & & & & & \\
 & 1 & & & 0 & \\
 & & 1 & & & \\
\hdashline
0 & -(Z_i - Z_j) & (Y_i - Y_j) & 1 & & \\
(Z_i - Z_j) & 0 & -(X_i - X_j) & & 1 & \\
-(Y_i - Y_j) & (X_i - X_j) & 0 & & & 1
\end{array}\right] \tag{8.6}$$

$\mathbf{H}_{ji}$ will be called a *translation matrix* as its action is to take forces at i and translate their effects to point j: it may be thought of as a set of force vectors at j due to the application of unit forces at i.

It can be seen from fig. 8.5(b) that the forces on members (m) and (n) on either side of joint j are equal and opposite; that is, that

$$\mathbf{P}_{ji} = -\mathbf{P}_{jk}$$

Hence from (8.5) we can write

$$\mathbf{P}_{ji} = -\mathbf{H}_{ji}\mathbf{P}_{ij} \tag{8.7}$$

This equation expresses the condition of equilibrium of member (*m*); and for this reason translation matrices are sometimes known as *equilibrium matrices*. The equilibrium condition of the member could also be written

$$\mathbf{P}_{ij} = \mathbf{H}_{ij}\mathbf{P}_{ji}$$

because of the definition of a translation matrix, and from this it can be seen that translation matrices have the curious property that

$$\mathbf{H}_{ij} = \mathbf{H}_{ji}^{-1} \tag{8.8}$$

Thus such a matrix could be inverted by changing the sign of the off-diagonal terms.

Let us now assume that member (*m*) is rigid, and that we want to know the displacement system $\mathbf{\Delta}_i$ at *i* due to a displacement $\mathbf{\Delta}_j$ of joint *j*. For equilibrium of (*m*) the virtual work done by the force system acting through the displacement system must be zero. As a rigid-body displacement is being given to (*m*), the internal forces do no work and the virtual work equation becomes

$$\delta W = \mathbf{P}_{ij}^T\mathbf{\Delta}_i + \mathbf{P}_{ji}^T\mathbf{\Delta}_j = 0$$

From (8.7)

$$\mathbf{P}_{ij}^T\mathbf{\Delta}_i = \mathbf{P}_{ij}^T\mathbf{H}_{ji}^T\mathbf{\Delta}_j$$

and as this is true for any values of $\mathbf{P}_{ij}$, we may write

$$\mathbf{\Delta}_i = \mathbf{H}_{ji}^T\mathbf{\Delta}_j \tag{8.9}$$

which is the translation relationship between displacements.

The complementary nature of the translation transformations of force (8.5) and displacement (8.9) is known as *contragredience*: it is a principle which will reappear later in other forms.

An analogous expression to the equilibrium equation (8.7) may be obtained relating the member deformations $\mathbf{\Delta}_{ij}$ and $\mathbf{\Delta}_{ji}$ at each end of a member. (It will be remembered that $\mathbf{\Delta}_{ij}$, as it has a double subscript, is the *relative* displacement of *i* relative to *j*.) Fig. 8.6(a) represents a member fixed at *i* with a set of displacements $\mathbf{\Delta}_j$ at *j*. Evidently in this case $\mathbf{\Delta}_j = \mathbf{\Delta}_{ji}$. With the member in this deformed state the deformation of *i* relative to *j* may be obtained by considering the deflection of the member to be split up into two components: a rigid-body displacement (fig. 8.6(b)) and a displacement of *i* relative to *j* (fig. 8.6(c)). From this it can be seen that

$$\mathbf{\Delta}_{ij} = -\mathbf{H}_{ji}^T\mathbf{\Delta}_{ji} \tag{8.10}$$

Let us now suppose that the flexibility and stiffness matrices of end j of member (n) relative to the fixed end at k are $\mathbf{F}^n_{ji}$ and $\mathbf{K}^n_{ji}$: the superscripts on the matrices refer to the member to which they pertain and the subscripts are used in the sense defined for flexibility and stiffness coefficients in Chapter 4. Now, for this particular member,

$$\mathbf{F}^n_{ji}\mathbf{P}_{jk} = \mathbf{\Delta}_{jk} = \mathbf{\Delta}_j$$

From (8.5) and (8.9) this could be written

$$\mathbf{H}^T_{ji}\mathbf{F}^n_{jj}\mathbf{H}_{ji}\mathbf{P}_{ij} = \mathbf{\Delta}_i$$

or

$$\mathbf{F}^n_{ii}\mathbf{P}_{ij} = \mathbf{\Delta}_i$$

where

$$\mathbf{F}^n_{ii} = \mathbf{H}^T_{ji}\mathbf{F}^n_{jj}\mathbf{H}_{ji} \tag{8.11}$$

The flexibility matrix $\mathbf{F}^n_{ii}$ represents the displacement at i as a result of unit forces applied at i, due solely to the elastic deformations of member (n). Thus

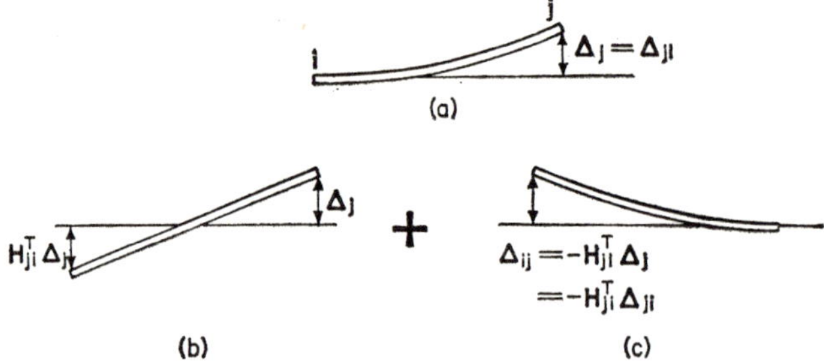

FIG. 8.6 Displacement and deformation relationships

in equation (8.11) the translation matrix $\mathbf{H}_{ji}$ has been used to transfer the point of applicability of a flexibility matrix from point j to point i: $\mathbf{H}_{ji}$ applied a translation transformation to $\mathbf{F}^n_{jj}$. The stiffness matrix $\mathbf{K}^n_{jj}$ can be shown to transform in a somewhat similar way, so that assuming (m) is a rigid extension on the end of (n), then

$$\mathbf{K}^n_{ii} = \mathbf{H}_{ij}\mathbf{K}^n_{jj}\mathbf{H}^T_{ij} \tag{8.12}$$

The total flexibility of point i is a combination of the deformation effects of both members. The combined flexibility matrix is

$$\begin{aligned}\mathbf{F}_{ii} &= \mathbf{F}^m_{ii} + \mathbf{F}^n_{ii} \\ &= \mathbf{F}^m_{ii} + \mathbf{H}^T_{ji}\mathbf{F}^n_{jj}\mathbf{H}_{ji}\end{aligned} \tag{8.13}$$

and if, in general, external forces are applied to both i and j, the resulting deflections are given by

$$\begin{Bmatrix} \boldsymbol{\Delta}_i \\ \boldsymbol{\Delta}_j \end{Bmatrix} = \left[\begin{array}{c|c} \mathbf{F}_{ii} & \mathbf{F}_{ij} \\ \hline \mathbf{F}_{ji} & \mathbf{F}_{jj} \end{array}\right] \begin{Bmatrix} \mathbf{P}_i \\ \mathbf{P}_j \end{Bmatrix}$$

$$= \left[\begin{array}{c|c} \mathbf{F}_{ii}^m + \mathbf{H}_{ji}^T\mathbf{F}_{jj}^n\mathbf{H}_{ji} & \mathbf{H}_{ji}^T\mathbf{F}_{ij}^n \\ \hline \mathbf{F}_{jj}^n\mathbf{H}_{ji} & \mathbf{F}_{jj}^n \end{array}\right] \begin{Bmatrix} \mathbf{P}_i \\ \mathbf{P}_j \end{Bmatrix} \tag{8.14}$$

Flexibility effects can be added in this way for all structures with a tree topology in which the members are in series with one another. This point will be elaborated in Section 8.8.

Stiffness matrices may not be combined in the same way, however. Broadly speaking, whereas flexibility matrices may be summed for structural elements in series, stiffness matrices can only be added for elements in parallel. As explained in Chapter 4, stiffness matrices can be formulated for structures which are kinematically indeterminate as well as for those which are determinate, a property not possessed by flexibility matrices. Hence a stiffness matrix may be formulated for an independent structural element such as (m) in fig. 8.5(b) relating the forces and displacements at both its ends. For this particular element,

$$\begin{Bmatrix} \mathbf{P}_{ij} \\ \mathbf{P}_{ji} \end{Bmatrix} = \begin{bmatrix} \mathbf{K}_{ii}^m & \mathbf{K}_{ij}^m \\ \mathbf{K}_{ji}^m & \mathbf{K}_{jj}^m \end{bmatrix} \begin{bmatrix} \boldsymbol{\Delta}_i \\ \boldsymbol{\Delta}_j \end{bmatrix} \tag{8.15}$$

$$= \mathbf{K}^m\boldsymbol{\Delta}$$

$\mathbf{K}^m$ is called the *complete stiffness matrix* of element (m). A good deal of redundant information is contained in the matrix as enough information to describe the entire stiffness characteristics of the member is contained in the submatrix $\mathbf{K}_{ii}^m$. In fact, using the equilibrium relation (8.7) and the member deformation relation (8.10), the complete stiffness matrix can be written in terms of $\mathbf{K}_{ii}^m$ as

$$\mathbf{K}^m = \left[\begin{array}{c|c} \mathbf{K}_{ii}^m & -\mathbf{K}_{ii}^m\mathbf{H}_{ji}^T \\ \hline -\mathbf{H}_{ji}\mathbf{K}_{ii}^m & \mathbf{H}_{ji}\mathbf{K}_{ii}^m\mathbf{H}_{ji}^T \end{array}\right]$$

$$= \left[\begin{array}{c} \mathbf{I} \\ \hline -\mathbf{H}_{ji} \end{array}\right] [\mathbf{K}_{ii}^m] [\mathbf{I} \mid -\mathbf{H}_{ji}^T] \tag{8.16}$$

It will be seen in Section 8.6 that although a complete stiffness matrix contains redundant information and appears to be a rather clumsy concept, its use leads to a particularly neat formulation of the stiffness matrix of a structure as a whole. The complete stiffness matrix for a member is of course singular in that no corresponding flexibility matrix could be formed to relate forces and displacements at either end of a member.

8.5 Rotation Matrices

The previous section showed how the effects of forces and displacements could be translated from place to place, but the effects always maintained the same orientation. In order to obtain force and displacement components in system coordinates when their components are already known in member coordinates, and *vice versa*, a new set of transformation matrices must be developed.

Considering first the transformation of force vectors, then we may write

$$\left.\begin{aligned} \mathbf{p}_i &= \mathbf{T}_{ij}\mathbf{P}_i \\ \text{and} \qquad \mathbf{p}_{ij} &= \mathbf{T}_{ij}\mathbf{P}_{ij} \end{aligned}\right\} \tag{8.17}$$

Where $\mathbf{T}_{ij}$ is called a rotation transformation matrix, or simply a *rotation matrix*: it is given by

$$\mathbf{T}_{ij} = \left[\begin{array}{ccc|ccc} \overline{xX} & \overline{xY} & \overline{xZ} & & & \\ \overline{yX} & \overline{yY} & \overline{yZ} & & 0 & \\ \overline{zX} & \overline{zY} & \overline{zZ} & & & \\ \hline & & & \overline{xX} & \overline{xY} & \overline{xZ} \\ & 0 & & \overline{yX} & \overline{yY} & \overline{yZ} \\ & & & \overline{zX} & \overline{zY} & \overline{zZ} \end{array}\right] \tag{8.18}$$

where, for instance, $\overline{xY}$ is the cosine of the angle between the x and Y axes and the other elements are similarly the cosines between the specified pairs of axes. Thus $\mathbf{T}_{ij}$ is a matrix of the direction cosines of the member axes of element *i–j* with regard to the system axes.

The displacements at a point are also described by a vector and so must transform according to the same transformation law in rectangular co-ordinates. Thus

$$\left.\begin{aligned} \boldsymbol{\delta}_i &= \mathbf{T}_{ij}\boldsymbol{\Delta}_i \\ \text{and} \qquad \boldsymbol{\delta}_{ij} &= \mathbf{T}_{ij}\boldsymbol{\Delta}_{ij} \end{aligned}\right\} \tag{8.19}$$

The work done by the product of the force and displacement vectors at a point must be independent of the coordinates in which they are expressed, so

$$\mathbf{p}_i^T \boldsymbol{\delta}_i = \mathbf{P}_i^T \boldsymbol{\Delta}_i$$

Substituting from (8.17) we get

$$\mathbf{P}_i^T \mathbf{T}_{ij}^T \boldsymbol{\delta}_i = \mathbf{P}_i^T \boldsymbol{\Delta}_i$$

and as this is true for any value of $\mathbf{P}_i$ we can write

$$\mathbf{T}_{ij}^T \boldsymbol{\delta}_i = \boldsymbol{\Delta}_i \tag{8.20}$$

which is the inverse transformation of (8.19). From (8.19) and (8.20) we can see that

$$\mathbf{T}_{ij}^T = \mathbf{T}_{ij}^{-1} \tag{8.21}$$

or

$$\mathbf{T}_{ij}^T \mathbf{T}_{ij} = \mathbf{I} \tag{8.22}$$

Thus the inverse of a rotation transformation matrix is equal to its transpose: a matrix with this property is called an *orthogonal transformation matrix*.

From equations (8.17), (8.19) and (8.21) we can readily see that the relations transforming the flexibility and stiffness matrices of end i of member (m) from member to system coordinates are

$$\left.\begin{aligned} \mathbf{F}_{ii}^m &= \mathbf{T}_{ij}^T \mathbf{f}_{ii}^m \mathbf{T}_{ij} \\ \mathbf{K}_{ii}^m &= \mathbf{T}_{ij}^T \mathbf{k}_{ii}^m \mathbf{T}_{ij} \end{aligned}\right\} \tag{8.23}$$

It is interesting to note that the rotation transformation is the same for both flexibility and stiffness matrices.

When both rotation and translation transformations have to be carried out, the order in which the transformations are applied is immaterial as long as the translation matrices are written with respect to the correct coordinate system at the time of application. If, for instance, the complete stiffness matrix of some element (m) connecting nodes i and j is required in system coordinates and $\mathbf{k}_{ii}^m$ is known in member coordinates, then we could proceed along two alternative courses: we could either initially form the complete stiffness matrix in member coordinates,

$$\mathbf{k}^m = \left[\begin{array}{c} \mathbf{I} \\ \hline -\mathbf{h}_{ji} \end{array}\right] [\mathbf{k}_{ii}^m] \left[\mathbf{I} \,\middle|\, -\mathbf{h}_{ji}^T\right]$$

$$= \left[\begin{array}{c|c} \mathbf{k}_{ii}^m & -\mathbf{k}_{ii}^m \mathbf{h}_{ji}^T \\ \hline -\mathbf{h}_{ji}\mathbf{k}_{ii}^m & \mathbf{h}_{ji}\mathbf{k}_{ii}^m \mathbf{h}_{ji}^T \end{array}\right]$$

in which the translation matrix $\mathbf{h}_{ij}$ is expressed in member coordinates and then apply a rotation transformation to obtain the required result:

$$\mathbf{K}^m = \begin{bmatrix} \mathbf{T}_{ij}^T & 0 \\ 0 & \mathbf{T}_{ij}^T \end{bmatrix} \mathbf{k}^m \begin{bmatrix} \mathbf{T}_{ij} & 0 \\ 0 & \mathbf{T}_{ij} \end{bmatrix}$$

or we could begin by rotating $\mathbf{k}_{ii}^m$ into system coordinates:

$$\mathbf{K}_{ii}^m = \mathbf{T}_{ij}^T \mathbf{k}_{ii}^m \mathbf{T}_{ij}$$

and continue by obtaining the complete stiffness matrix with the use of a system-coordinate translation matrix $\mathbf{H}_{ji}$, thus

$$\mathbf{K}^m = \begin{bmatrix} \mathbf{I} \\ -\mathbf{H}_{ji} \end{bmatrix} [\mathbf{K}_{ii}^m] [\mathbf{I} \mid -\mathbf{H}_{ji}^T]$$

as in equation (8.16). The second approach would evidently involve less matrix multiplication.

The form of the **T**-matrix can sometimes be complicated, but it is usually

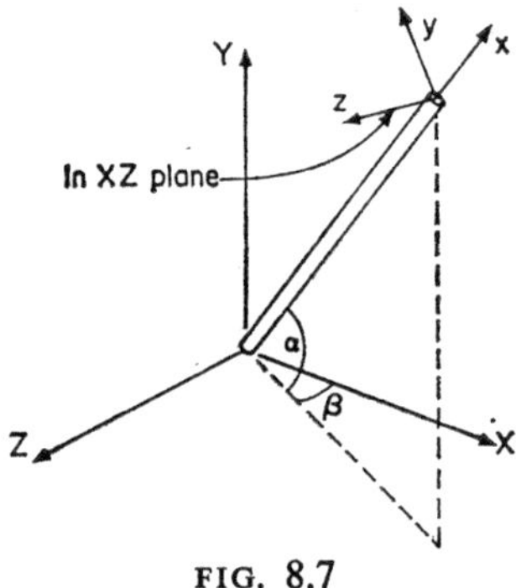

FIG. 8.7

fairly simple. For the three-dimensional situation of fig. 8.7, the rotation matrix is

$$\mathbf{T}_{ij} = \left[\begin{array}{ccc|ccc} \cos\alpha\cos\beta & \sin\alpha & \cos\alpha\sin\beta & & & \\ -\sin\alpha\cos\beta & \cos\alpha & -\sin\alpha\sin\beta & & 0 & \\ -\sin\beta & 0 & \cos\beta & & & \\ \hline & & & \cos\alpha\cos\beta & \sin\alpha & \cos\alpha\sin\beta \\ & 0 & & -\sin\alpha\cos\beta & \cos\alpha & -\sin\alpha\sin\beta \\ & & & -\sin\beta & 0 & \cos\beta \end{array}\right] \tag{8.24}$$

For either of the two-dimensional situations shown in fig. 8.8, $\beta = 0$ in equation (8.24) and the resulting contracted **T**-matrix relating

$$\mathbf{p}_{ij} = \begin{Bmatrix} p_{ij} \\ q_{ij} \\ n_{ij} \end{Bmatrix}$$

$$\mathbf{P}_{ij} = \begin{Bmatrix} P_{ij} \\ Q_{ij} \\ N_{ij} \end{Bmatrix}$$

is

$$\mathbf{T}_{ij} = \begin{bmatrix} \cos\alpha & \sin\alpha & 0 \\ -\sin\alpha & \cos\alpha & 0 \\ 0 & 0 & 1 \end{bmatrix} \tag{8.25}$$

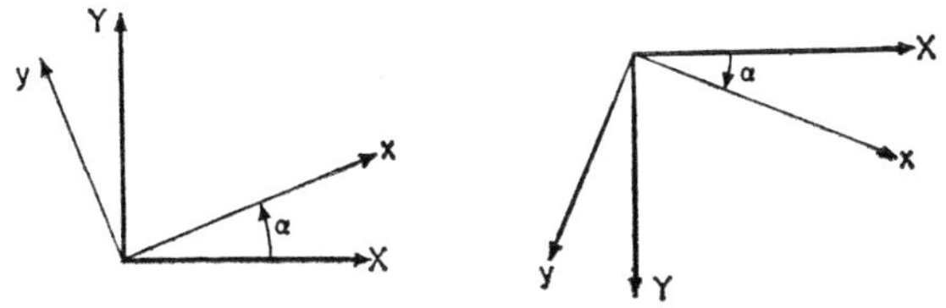

FIG. 8.8

8.6 The Matrix Displacement (Stiffness) Method

The essence of the displacement method for the analysis of a structure is to form a relationship

$$\mathbf{P} = \mathbf{K}\boldsymbol{\Delta} \tag{8.26}$$

between all the displacement unknowns $\boldsymbol{\Delta}$ of the structure and the corresponding external forces **P** applied to the nodes: **K** is, of course, the stiffness matrix of the structure. Once such an equation has been obtained, it can be solved for the nodal displacements $\boldsymbol{\Delta}$, and from these the member forces may be found. The main problem, then, is to obtain **K** from the various member stiffness matrices of the structure.

The two assumptions made at this stage (apart from the normal ones of linear analysis) are:

(a) The only external forces applied to the structure are applied at its nodes. If forces are applied at intermediate points, they should be replaced by statically equivalent nodal forces as explained in Section 3.5.

(b) Enough members are attached to rigid foundations for the structure to be stable as a rigid body. For civil engineering structures this will always be true, but some marine and aerospace structures are not attached to any rigid reference frame so that an artificial frame has to be introduced. We shall not concern ourselves with this problem in this book.

Now, for any member (r) joining nodes i and j we can write

$$\mathbf{P}^r = \mathbf{K}^r\mathbf{\Delta}^r \tag{8.27}$$

where

$$\mathbf{P}^r = \begin{Bmatrix} \mathbf{P}_{ij} \\ --- \\ \mathbf{P}_{ji} \end{Bmatrix}$$

$$\mathbf{\Delta}^r = \begin{Bmatrix} \mathbf{\Delta}_i \\ --- \\ \mathbf{\Delta}_j \end{Bmatrix}$$

and where $\mathbf{K}^r$ is the complete stiffness matrix of the element. For the m members of the structure, the m equations exemplified by equation (8.27) can be assembled into the form

$$\begin{Bmatrix} \mathbf{P}^1 \\ \mathbf{P}^2 \\ \vdots \\ \mathbf{P}^m \end{Bmatrix} = \begin{bmatrix} \mathbf{K}^1 & & & \\ & \mathbf{K}^2 & & \\ & & \ddots & \\ & & & \mathbf{K}^m \end{bmatrix} \begin{Bmatrix} \mathbf{\Delta}^1 \\ \mathbf{\Delta}^2 \\ \vdots \\ \mathbf{\Delta}^m \end{Bmatrix} \tag{8.28}$$

which could be written simply as

$$\bar{\mathbf{P}} = \bar{\mathbf{K}}\bar{\mathbf{\Delta}} \tag{8.29}$$

This equation does not, of course, couple the effects of the different members together: it is merely an assemblage of individual member equations at this stage.

The relationship between the vectors of nodal displacements $\mathbf{\Delta}$ and of member-end displacements $\bar{\mathbf{\Delta}}$ could be determined quite simply from the geometry of the structure. In matrix form this could be written as

$$\bar{\mathbf{\Delta}} = \mathbf{A}\mathbf{\Delta} \tag{8.30}$$

where $\mathbf{A}$ is called the *connection matrix* of the structure (it connects the effects of individual members together to give the behaviour of the complete structure), or sometimes simply a *boolean matrix* as it consists entirely of ones and zeros if the member and system effects are in the same coordinates. Consider the structure of fig. 8.9, for example. The only nodal displacements will be $\boldsymbol{\Delta}_2$ and $\boldsymbol{\Delta}_3$, while each of the four members will have a set of end displacements for each end. Equation (8.30) becomes in this instance.

$$\begin{Bmatrix} \boldsymbol{\Delta}_1 \\ \boldsymbol{\Delta}_2 \\ \hline \boldsymbol{\Delta}_2 \\ \boldsymbol{\Delta}_3 \\ \hline \boldsymbol{\Delta}_1 \\ \boldsymbol{\Delta}_3 \\ \hline \boldsymbol{\Delta}_3 \\ \boldsymbol{\Delta}_4 \end{Bmatrix} = \begin{bmatrix} 0 & 0 \\ \mathbf{I} & 0 \\ \hline \mathbf{I} & 0 \\ 0 & \mathbf{I} \\ \hline 0 & 0 \\ 0 & \mathbf{I} \\ \hline 0 & \mathbf{I} \\ 0 & 0 \end{bmatrix} \begin{Bmatrix} \boldsymbol{\Delta}_2 \\ \hline \boldsymbol{\Delta}_3 \end{Bmatrix} \tag{8.31}$$

where **I** is a unit sub-matrix: a diagonal matrix of ones. The actual dimensions of **I** in each case will depend on how many degrees of freedom are considered at the nodes: six for a three-dimensional problem or three for a two-dimensional problem.

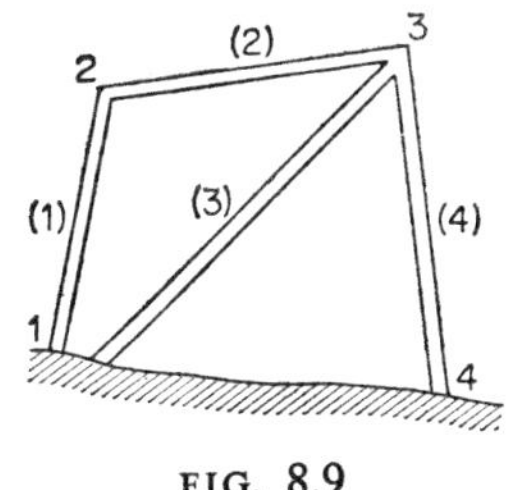

FIG. 8.9

The contragredient relationship between member and nodal forces may be obtained by the use of virtual work. If the structure is subjected to nodal forces **P** and has within it member forces $\overline{\mathbf{P}}$, then if a series of virtual displacements $\boldsymbol{\Delta}$ were applied leading to member displacements $\overline{\boldsymbol{\Delta}}$, then the internal and external work done by the forces acting through the displacements would be equal, and

$$\overline{\boldsymbol{\Delta}}^T\overline{\mathbf{P}} = \boldsymbol{\Delta}^T\mathbf{P}$$

From (8.30) this could be written

$$\boldsymbol{\Delta}^T\mathbf{A}^T\overline{\mathbf{P}} = \boldsymbol{\Delta}^T\mathbf{P}$$

and as this is true for any set of nodal displacements $\boldsymbol{\Delta}$, we may write

$$\mathbf{A}^T\overline{\mathbf{P}} = \mathbf{P} \tag{8.32}$$

Substituting (8.30) and (8.32) into (8.29), we then have the relationship

$$\mathbf{P} = \mathbf{A}^T\overline{\mathbf{K}}\mathbf{A}\boldsymbol{\Delta} \tag{8.33}$$

so that from (8.26) we can state that the stiffness matrix of the complete structure is given by

$$\mathbf{K} = \mathbf{A}^T\overline{\mathbf{K}}\mathbf{A} \tag{8.34}$$

Thus the function of the connection matrix is to assemble the stiffness matrices of the individual members into the overall stiffness matrix of the structure.

For the structure of fig. 8.9, the matrix of member stiffnesses will be

$$\overline{\mathbf{K}} = \begin{bmatrix} \mathbf{K}^1 & & & \\ & \mathbf{K}^2 & & \\ & & \mathbf{K}^3 & \\ & & & \mathbf{K}^4 \end{bmatrix}$$

$$= \left[\begin{array}{cc:cc:cc:cc} \mathbf{K}^1_{11} & \mathbf{K}^1_{12} & & & & & & \\ \mathbf{K}^1_{21} & \mathbf{K}^1_{22} & & & & & & \\ \hdashline & & \mathbf{K}^2_{22} & \mathbf{K}^2_{23} & & & & \\ & & \mathbf{K}^2_{32} & \mathbf{K}^2_{33} & & & & \\ \hdashline & & & & \mathbf{K}^3_{11} & \mathbf{K}^3_{13} & & \\ & & & & \mathbf{K}^3_{31} & \mathbf{K}^3_{33} & & \\ \hdashline & & & & & & \mathbf{K}^4_{33} & \mathbf{K}^4_{34} \\ & & & & & & \mathbf{K}^4_{43} & \mathbf{K}^4_{44} \end{array}\right]$$

Using the connection matrix of (8.31) in equation (8.34), the stiffness matrix of the structure then becomes

$$\mathbf{K} = \left[\begin{array}{c:c} \mathbf{K}^1_{22} + \mathbf{K}^2_{22} & \mathbf{K}^2_{23} \\ \hdashline \mathbf{K}^2_{32} & \mathbf{K}^2_{33} + \mathbf{K}^3_{33} + \mathbf{K}^4_{33} \end{array}\right] \tag{8.35}$$

Once $\mathbf{K}$ has been found, the nodal displacements of the structure may be obtained from equation (8.26). This could be written formally as

$$\mathbf{\Delta} = \mathbf{K}^{-1}\mathbf{P}$$

but in fact a stiffness matrix would seldom if ever actually be inverted: it is generally far simpler and more efficient merely to solve the set of equations represented by (8.26) directly. In any case, once $\boldsymbol{\Delta}$ is known then the member end forces are given by

$$\bar{\mathbf{P}} = \bar{\mathbf{K}}\mathbf{A}\boldsymbol{\Delta} \tag{8.36}$$

and if the member end forces are required in terms of member coordinates then they may be obtained by the transformation

$$\bar{\mathbf{p}} = \bar{\mathbf{T}}\bar{\mathbf{P}} \tag{8.37}$$

where $\mathbf{T}$ is a diagonally partitioned matrix of the rotation matrices of each member-end.

When specific displacements rather than forces are applied to some of the nodes of a structure due, for example, to foundation settlement or thermal effects, the formulation of the matrix displacement method becomes slightly more complex. Suppose that the vector of nodal displacements is partitioned into two sub-matrices in the form

$$\boldsymbol{\Delta} = \begin{Bmatrix} \boldsymbol{\Delta}_\alpha \\ \hline \boldsymbol{\Delta}_\beta \end{Bmatrix} \tag{8.38}$$

Where $\boldsymbol{\Delta}_\alpha$ are known, specified displacements at certain of the nodes in a structure and $\boldsymbol{\Delta}_\beta$ are the unknown displacements of the remaining degrees of freedom of the structure. The vector of applied forces can be partitioned in the same way to give

$$\mathbf{P} = \begin{Bmatrix} \mathbf{P}_\alpha \\ \hline \mathbf{P}_\beta \end{Bmatrix}$$

In this case, however, it is the forces $\mathbf{P}_\alpha$ which are unknown (they may be called *reactive forces*), while the forces $\mathbf{P}_\beta$ are known and specified (though the forces applied to many nodes may be zero). The connection matrix $\mathbf{A}$, too, can be partitioned so that (8.30) may be written

$$\begin{aligned} \bar{\boldsymbol{\Delta}} &= \mathbf{A}\boldsymbol{\Delta} \\ &= [\mathbf{A}_\alpha \mathbf{A}_\beta] \begin{Bmatrix} \boldsymbol{\Delta}_\alpha \\ \boldsymbol{\Delta}_\beta \end{Bmatrix} \\ &= \mathbf{A}_\alpha \boldsymbol{\Delta}_\alpha + \mathbf{A}_\beta \boldsymbol{\Delta}_\beta \end{aligned} \tag{8.39}$$

Similarly, the relationship between internal and external forces becomes

$$\begin{bmatrix} \mathbf{A}_\alpha^T \\ \mathbf{A}_\beta^T \end{bmatrix} \overline{\mathbf{P}} = \begin{Bmatrix} \mathbf{P}_\alpha \\ \mathbf{P}_\beta \end{Bmatrix} \tag{8.40a}$$

or

$$\begin{aligned} \mathbf{A}_\alpha^T \overline{\mathbf{P}} &= \mathbf{P}_\alpha \\ \mathbf{A}_\beta^T \overline{\mathbf{P}} &= \mathbf{P}_\beta \end{aligned} \tag{8.40b}$$

Hence equation (8.33) may now be written in the partitioned form

$$\begin{Bmatrix} \mathbf{P}_\alpha \\ \mathbf{P}_\beta \end{Bmatrix} = \begin{bmatrix} \mathbf{A}_\alpha^T \\ \mathbf{A}_\beta^T \end{bmatrix} \overline{\mathbf{K}} \, [\mathbf{A}_\alpha \mathbf{A}_\beta] \begin{Bmatrix} \mathbf{\Delta}_\alpha \\ \mathbf{\Delta}_\beta \end{Bmatrix}$$

which in turn may now be written as the two separate equations

$$\mathbf{P}_\alpha = \mathbf{A}_\alpha^T \overline{\mathbf{K}} \mathbf{A}_\alpha \mathbf{\Delta}_\alpha + \mathbf{A}_\alpha^T \overline{\mathbf{K}} \mathbf{A}_\beta \mathbf{\Delta}_\beta \tag{8.41}$$

$$\mathbf{P}_\beta = \mathbf{A}_\beta^T \overline{\mathbf{K}} \mathbf{A}_\alpha \mathbf{\Delta}_\alpha + \mathbf{A}_\beta^T \overline{\mathbf{K}} \mathbf{A}_\beta \mathbf{\Delta}_\beta \tag{8.42}$$

The second equation may be reformulated as

$$\mathbf{A}_\beta^T \overline{\mathbf{K}} \mathbf{A}_\beta \mathbf{\Delta}_\beta = \mathbf{P}_\beta - \mathbf{A}_\beta^T \overline{\mathbf{K}} \mathbf{A}_\alpha \mathbf{\Delta}_\alpha \tag{8.43}$$

The solution procedure is now as follows:

(a) Equation (8.43) is solved for $\mathbf{\Delta}_\beta$
(b) $\mathbf{P}_\alpha$ is found by substituting $\mathbf{\Delta}_\beta$ into (8.41), if the reactive forces are required.
(c) The member forces $\overline{\mathbf{P}}$ are found from (8.36).
(d) The member forces may be obtained in member coordinates from (8.37).

There should be no difficulty in finding the matrices $\mathbf{A}_\alpha$ and $\mathbf{A}_\beta$ as they are merely parts of the overall connection matrix $\mathbf{A}$, though the order of the columns of $\mathbf{A}$ might be altered. For example, if the X-displacement U_3 at node 3 and the Y-displacement V_2 at node 2 were specified for the structure of fig. 8.9, then

$$\mathbf{\Delta}_\alpha = \begin{Bmatrix} U_3 \\ V_2 \end{Bmatrix} \qquad \mathbf{\Delta}_\beta = \begin{Bmatrix} U_2 \\ \Psi_2 \\ V_3 \\ \Psi_3 \end{Bmatrix}$$

and equation (8.39) becomes in its fully expanded form

$$
\begin{Bmatrix} U_{12} \\ V_{12} \\ \Psi_{12} \\ \hline U_{21} \\ V_{21} \\ \Psi_{21} \\ \hline U_{23} \\ V_{23} \\ \Psi_{23} \\ \hline U_{32} \\ V_{32} \\ \Psi_{32} \\ \hline U_{31} \\ V_{31} \\ \Psi_{31} \\ \hline U_{13} \\ V_{13} \\ \Psi_{13} \\ \hline U_{34} \\ V_{34} \\ \Psi_{34} \\ \hline U_{43} \\ V_{43} \\ \Psi_{43} \end{Bmatrix}
=
\left[\begin{array}{cc|cccc}
0 & 0 & 0 & 0 & 0 & 0 \\
0 & 0 & 0 & 0 & 0 & 0 \\
0 & 0 & 0 & 0 & 0 & 0 \\
\hline
0 & 0 & 1 & 0 & 0 & 0 \\
0 & 1 & 0 & 0 & 0 & 0 \\
0 & 0 & 0 & 1 & 0 & 0 \\
\hline
0 & 0 & 1 & 0 & 0 & 0 \\
0 & 1 & 0 & 0 & 0 & 0 \\
0 & 0 & 0 & 1 & 0 & 0 \\
\hline
1 & 0 & 0 & 0 & 0 & 0 \\
0 & 0 & 0 & 0 & 1 & 0 \\
0 & 0 & 0 & 0 & 0 & 1 \\
\hline
1 & 0 & 0 & 0 & 0 & 0 \\
0 & 0 & 0 & 0 & 1 & 0 \\
0 & 0 & 0 & 0 & 0 & 1 \\
\hline
0 & 0 & 0 & 0 & 0 & 0 \\
0 & 0 & 0 & 0 & 0 & 0 \\
0 & 0 & 0 & 0 & 0 & 0 \\
\hline
1 & 0 & 0 & 0 & 0 & 0 \\
0 & 0 & 0 & 0 & 1 & 0 \\
0 & 0 & 0 & 0 & 0 & 1 \\
\hline
0 & 0 & 0 & 0 & 0 & 0 \\
0 & 0 & 0 & 0 & 0 & 0 \\
0 & 0 & 0 & 0 & 0 & 0
\end{array}\right]
\begin{Bmatrix} U_3 \\ V_2 \\ \hline U_2 \\ \Psi_2 \\ V_3 \\ \Psi_3 \end{Bmatrix}
\tag{8.44}
$$

The use of the connection matrix **A** in the assembly of the overall stiffness matrix of a structure is perfectly general. In the examples given here, all the elements of **A** have been either zero or unity. If, however, the member forces and displacements were in member coordinates while the nodal forces and displacements were in system coordinates, a connection matrix could still be formed relating them, but its elements would take other values besides 0 and 1. A connection matrix could also be obtained to form an overall stiffness matrix from member-end stiffness matrices of the members: this would halve the size of the connection matrix but would make its formulation more complex—it would for instance have to contain translation transformation sub-matrices in appropriate positions.

It is an interesting point that the connection matrix may also be used to obtain the member forces in a statically determinate structure subjected to a set of externally applied nodal forces. In general, equation (8.32) cannot be used to find $\bar{\mathbf{P}}$ from $\mathbf{P}$ as $\mathbf{A}$ is not a square matrix. If a structure is statically determinate, however, and $\bar{\mathbf{P}}$ consists of the minimum number of forces for each member to make its force distribution determinate (each member would have three terms for a two-dimensional structure and six for a three-dimensional structure) then $\mathbf{A}^T$ will be square and non-singular so that the expression

$$\bar{\mathbf{P}} = [\mathbf{A}^T]^{-1}\mathbf{P} \tag{8.45}$$

solves the statics of the structure.

Although the use of the concept of the connection matrix leads to a completely formal way of assembling member stiffness matrices into an overall stiffness matrix (equation (8.34)), connection matrices are seldom if ever used in practice. Instead, the overall stiffness matrix is assembled directly from the sub-matrices of the member stiffness in a manner which has already, in effect, been demonstrated for the structure of fig. 8.9 in the formation of equation (8.35). The procedure is as follows. The total stiffness matrix **K** may be partitioned into a number of sub-matrices such as $\mathbf{K}_{ij}$ which would give the forces at node i due to displacements at j. For the structure of fig. 8.9, the total matrix would simply be

$$\mathbf{K} = \left[\begin{array}{c|c} \mathbf{K}_{22} & \mathbf{K}_{23} \\ \mathbf{K}_{32} & \mathbf{K}_{33} \end{array}\right]$$

From the definition of a stiffness matrix, the diagonal sub-matrix $\mathbf{K}_{ii}$ relating the forces and displacements at node i of a structure must be the sum of the member-end stiffness matrices for end i of all members connected to node i, or

$$\mathbf{K}_{ii} = \sum_m \mathbf{K}_{ii}^m \tag{8.46}$$

If a member (r) joins nodes i and j then a set of displacements $\mathbf{\Delta}_i$ at i will cause forces

$$\mathbf{P}_{ji} = \mathbf{K}^r_{ji}\mathbf{\Delta}_i$$

to act on end j of (r), so that to restrain node j from moving an external force

$$\mathbf{P}_j = \mathbf{P}_{ji}$$

must be applied. Thus it is clear that a typical off-diagonal sub-matrix of $\mathbf{K}$ must be given by

$$\mathbf{K}_{ji} = \mathbf{K}^r_{ji} \tag{8.47}$$

where member (r) connects i and j. If the two nodes are not connected, then $\mathbf{K}_{ji}$ is zero, from which it can be seen that for a reasonably large structure, a large part of $\mathbf{K}$ will consist of zero elements.

This method of assembling the total stiffness matrix of a structure is known as the *direct stiffness method.*

8.7 Computational Points

The present chapter is written primarily on the assumption that the reader will at some time make use of a computer for structural analysis, and that in doing so he will use a standard, commercially available program: it is most unlikely that he will have to program the computer himself. There are, however, a few points about computational procedure which the casual computer user should be aware of, firstly because they may lead to a more efficient use of computer time, and secondly because they may lead to a greater awareness of possible inaccuracies in computer results.

The first point concerns the shape of the $\mathbf{K}$-matrix of a structure. For the frame structure of fig. 8.10(a), the stiffness matrix will be of the form shown on page 212 (equation (8.48)).

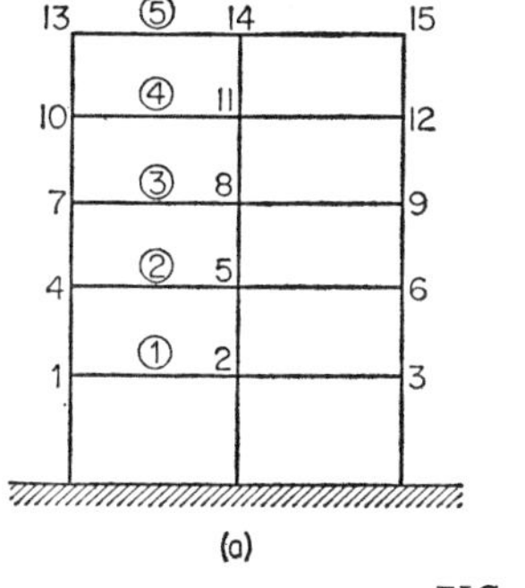

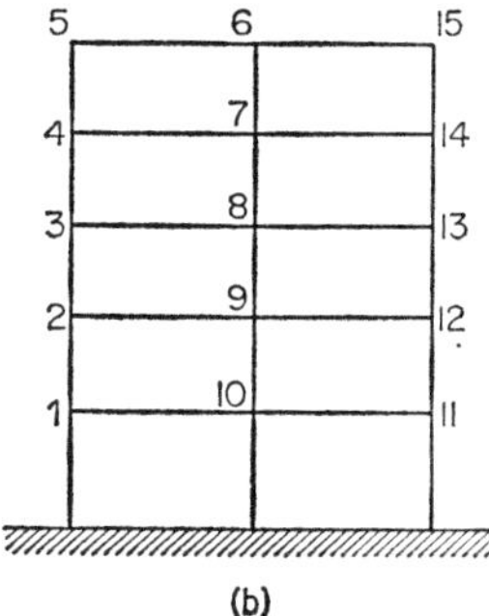

FIG. 8.10

$$\mathbf{K} = \left[\begin{array}{ccc|ccc|ccc|ccc|ccc}
\mathbf{K}_{11} & \mathbf{K}_{12} & & \mathbf{K}_{14} & & & & & & & & & & & \\
\mathbf{K}_{21} & \mathbf{K}_{22} & \mathbf{K}_{23} & & \mathbf{K}_{25} & & & & & & & & & & \\
 & \mathbf{K}_{32} & \mathbf{K}_{33} & & & \mathbf{K}_{36} & & & & & & & & & \\
\hline
\mathbf{K}_{41} & & & \mathbf{K}_{44} & \mathbf{K}_{45} & & \mathbf{K}_{47} & & & & & & & & \\
 & \mathbf{K}_{52} & & \mathbf{K}_{54} & \mathbf{K}_{55} & \mathbf{K}_{56} & & \mathbf{K}_{58} & & & & & & & \\
 & & \mathbf{K}_{63} & & \mathbf{K}_{65} & \mathbf{K}_{66} & & & \mathbf{K}_{69} & & & & & & \\
\hline
 & & & \mathbf{K}_{74} & & & \mathbf{K}_{77} & \mathbf{K}_{78} & & \mathbf{K}_{7,10} & & & & & \\
 & & & & \mathbf{K}_{85} & & \mathbf{K}_{87} & \mathbf{K}_{88} & \mathbf{K}_{89} & & \mathbf{K}_{8,11} & & & & \\
 & & & & & \mathbf{K}_{96} & & \mathbf{K}_{98} & \mathbf{K}_{99} & & & \mathbf{K}_{9,12} & & & \\
\hline
 & & & & & & \mathbf{K}_{10,7} & & & \mathbf{K}_{10,10} & \mathbf{K}_{10,11} & & \mathbf{K}_{10,13} & & \\
 & & & & & & & \mathbf{K}_{11,8} & & \mathbf{K}_{11,10} & \mathbf{K}_{11\ 11} & \mathbf{K}_{11,12} & & \mathbf{K}_{11,14} & \\
 & & & & & & & & \mathbf{K}_{12,9} & & \mathbf{K}_{12,11} & \mathbf{K}_{12,12} & & & \mathbf{K}_{12,15} \\
\hline
 & & & & & & & & & \mathbf{K}_{13,10} & & & \mathbf{K}_{13,13} & \mathbf{K}_{13,14} & \\
 & & & & & & & & & & \mathbf{K}_{14,11} & & \mathbf{K}_{14,13} & \mathbf{K}_{14,14} & \mathbf{K}_{14,15} \\
 & & & & & & & & & & & \mathbf{K}_{15,12} & & \mathbf{K}_{15,14} & \mathbf{K}_{15,15}
\end{array}\right] \tag{8.48}$$

This matrix has a regular pattern. A matrix of this form in which the non-zero elements in any row lie within a certain distance of the diagonal element is called a *banded matrix*. If for the five storeys of the structure $\mathbf{C}_{11}$ represents the stiffness relationship between forces and displacements at nodes on storey 1, $\mathbf{C}_{21}$ represents the relationships between forces on the second storey and displacements in the first, and so on, then $\mathbf{K}$ could also be written

$$\mathbf{K} = \begin{bmatrix} \mathbf{C}_{11} & \mathbf{C}_{12} & & & \\ \mathbf{C}_{21} & \mathbf{C}_{22} & \mathbf{C}_{23} & & \\ & \mathbf{C}_{32} & \mathbf{C}_{33} & \mathbf{C}_{34} & \\ & & \mathbf{C}_{43} & \mathbf{C}_{44} & \mathbf{C}_{45} \\ & & & \mathbf{C}_{54} & \mathbf{C}_{55} \end{bmatrix} \tag{8.49}$$

Each row of sub-matrices except the first and the last contains three terms: a matrix with this characteristic is called a *tridiagonal matrix*. It is interesting to note that even the non-zero sub-matrices $\mathbf{C}_{ij}$ themselves contain a great many zero terms. The sub-matrices on the diagonal are banded, and the off-diagonal sub-matrices are diagonally partitioned.

The banded nature of the stiffness matrix $\mathbf{K}$ is taken into account in structural analysis computer programs, and the efficiency of such a program depends to a considerable degree on the band width of the stiffness matrix: the narrower the band, the faster the calculation. The band width for any structure depends on the order in which the nodes are considered, certain choices in the order of numbering the nodes being more efficient than others. For example, if the structure of fig. 8.10 were numbered as in (b) rather than as in (a), the stiffness matrix would take the form shown on page 214 (equation (8.50)).

Comparing equations (8.48) and (8.50), it is evident that the numbering of fig. 8.10(a) is far more satisfactory than that of fig. 8.10(b), for in the former case the band width is 7 nodes while in the latter the band is as much as 19 nodes wide. In fact, the most efficient numbering scheme for a structure with a regular grid of nodes is to number the nodes along the rows of the grid in the shorter direction, with one row following after the next. The maximum difference between nodal numbers will then be equal to the number of nodes in a storey. This system is followed in fig. 8.10(a).

In view of the banded nature of the stiffness matrix of a large structure it would seem that the equations to be solved in a displacement method analysis will always be well-conditioned. Unfortunately this is not so in practice, for the conditioning of the problem deteriorates whenever stiffness coefficients of greatly differing magnitude occur in close proximity. Such a situation will

$$
\mathbf{K} = \begin{bmatrix}
K_{11} & K_{12} & & & & & & & & K_{1,10} & & & & & \\
K_{21} & K_{22} & K_{23} & & & & & & K_{29} & & & & & & \\
 & K_{32} & K_{33} & K_{34} & & & & K_{38} & & & & & & & \\
 & & K_{43} & K_{44} & K_{45} & & K_{47} & & & & & & & & \\
 & & & K_{54} & K_{55} & K_{56} & & & & & & & & & \\
 & & & & K_{65} & K_{66} & K_{67} & & & & & & & & K_{6,15} \\
 & & & K_{74} & & K_{76} & K_{77} & K_{78} & & & & & & K_{7,14} & \\
 & & K_{83} & & & & K_{87} & K_{88} & K_{89} & & & & K_{8,13} & & \\
 & K_{92} & & & & & & K_{98} & K_{99} & K_{9,10} & & K_{9,12} & & & \\
K_{10,1} & & & & & & & & K_{10,9} & K_{10,10} & K_{10,11} & & & & \\
 & & & & & & & & & K_{11,10} & K_{11,11} & K_{11,12} & & & \\
 & & & & & & & & K_{12,9} & & K_{12,11} & K_{12,12} & K_{12,13} & & \\
 & & & & & & & K_{13,8} & & & & K_{13,12} & K_{13,13} & K_{13,14} & \\
 & & & & & & K_{14,7} & & & & & & K_{14,13} & K_{14,14} & K_{14,15} \\
 & & & & & K_{15,6} & & & & & & & & K_{15,14} & K_{15,15}
\end{bmatrix} \tag{8.50}
$$

always arise when frame structures are being analysed with all degrees of freedom considered independently at each node, for the axial stiffness of a member will be very much greater than its bending and transverse stiffnesses: it also tends to arise in continuum problems when very small (and hence stiff) finite elements are intermingled with large elements. For such a continuum problem careful idealisation of the structure will improve accuracy, but frame structures have no choice of idealisation. However, frame analysis programs are available which neglect axial flexibility, certainly of beams and also sometimes of columns, though column axial deflections do significantly affect the dynamic behaviour of very tall buildings.

Whether the equation system is well conditioned or not, a computer program will still give results; but the results for an ill-conditioned structure analysed on a computer normally working with only a small number of significant figures may be seriously in error, quite apart from errors due to mistakes in data preparation. With a combination of an ill-conditioned structure and a small number of significant figures, errors of the order of 200% have been found, and even with more normal structures errors of the order of 15% are sometimes common. All results must therefore be checked carefully to see whether conditions of local and overall equilibrium are fulfilled. If there is some doubt about this and if the program allows it, it is often helpful to find the residual vector

$$\mathbf{R} = \mathbf{K}\boldsymbol{\Delta}' - \mathbf{P} \tag{8.51}$$

where $\boldsymbol{\Delta}' = \boldsymbol{\Delta} + \mathrm{d}\boldsymbol{\Delta}$ is the result actually obtained from the solution of equation (8.26) and $\boldsymbol{\Delta}$ is the true result: $\mathbf{R}$ should of course ideally be zero, but if it is substantially different then inaccuracies have arisen due to ill-conditioning of the stiffness matrix. In this case, substitution of equation (8.26) into (8.51) leaves the equation

$$\mathbf{K}\,\mathrm{d}\boldsymbol{\Delta} = \mathbf{R}$$

In other words, the residual vector is reintroduced into (8.26) as the load vector and from this equation a set of correction displacements $\mathrm{d}\boldsymbol{\Delta}$ may be found, in much the same way as approximate values of redundant forces were used in Section 5.4 to improve the accuracy of a force method. This process could be continued as an iterative process (known as *iterating on the residuals*) till a satisfactorily small residual vector is obtained. There are, of course, many other approaches to the problem of producing valid results for ill-conditioned structures, but these are beyond the scope of this book.

The most common fault when an engineer first begins to use a computer for structural analysis is that he approaches the problem with insufficient caution, believing, perhaps, that as long as data are provided in the correct format the results will also be correct. But it is far from true that a computer relegates an engineer to the status of a technician. It has already been shown that the form of the idealisation of the problem can affect the results considerably, and in any case errors in data preparation can sometimes occur. It is therefore important that the results of a computer program should always be examined critically before they are accepted, and if possible they should be compared with an independent approximate check. In any case, whether or not a computer analysis is used, the engineer is still faced with the often difficult task of reducing an actual building structure to a two-dimensional conceptual model: this can only be done by experience and good judgement.

8.8 The Matrix Force (Flexibility) Method

There are far fewer matrix analysis computer programs available based on the force method rather than on the displacement method, mainly because it is so much easier to systematise the matrix-displacement method. However, the force method does have certain advantages in that its use may be more efficient for structures with few redundants, and because its very lack of systematisation means that the analyst is free to choose whichever statically determinate system he prefers; for with the possibility of choice he can then avoid ill-conditioned systems.

The matrix-force method can be very simple if the redundants are so chosen that the statically determinate structure has a *tree* form of topology: such a form is attained for the structure of fig. 8.11(a), for example, if the statically

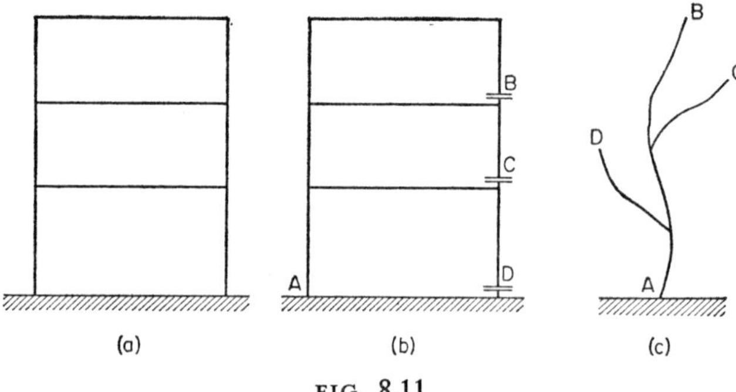

FIG. 8.11

determinate structure is produced by inserting complete cuts at the three points indicated in fig. 8.11(b). The paths between the built-in end A and points B, C and D are topologically similar in fig. 8.11(b and c)—hence the name 'tree topology'. The analysis proceeds in the general manner of force-method analyses by finding the flexibility relationships linking forces and relative displacements at the cuts—expressed in terms of a flexibility matrix—and by finding the relative displacements at the cuts due to applied loads. The redundant forces are then found by solving a set of compatibility equations, and from these the entire force distribution in the structure is found by statics. The particular advantage of a tree-type of statically determinate structure is that it enables the overall flexibility matrix to be found particularly easily by the summation procedure for members in series already given in equation (8.14). This is best illustrated by an example.

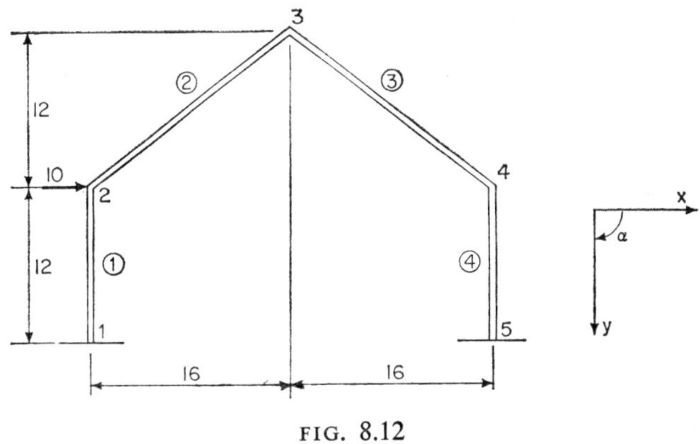

FIG. 8.12

The two-dimensional hipped portal frame of fig. 8.12 will be analysed by inserting a complete cut at point 5. The effect of axial deformations is assumed to be small and hence the flexibility of each member in the x-direction (the member coordinates used are those in fig. 8.4(b)) is zero. The section stiffness EI is the same for all members, so it will be taken to be unity.

The first step is to obtain the member flexibility matrices in system coordinates. To do this, the rotation transformation matrix of equation (8.25) is used: for the various members of the structure the angle α is:

Member	α
2–1	-90°
3–2	$-\tan^{-1}(\frac{12}{16})$
4–3	$+\tan^{-1}(\frac{12}{16})$
5–4	$+90^\circ$

For member (1) the flexibility matrix in member coordinates will be, using the terminology of Section 8.4,

$$\mathbf{f}^1_{22} = \begin{bmatrix} 0 & 0 & 0 \\ 0 & 12^3/3 & 12^2/2 \\ 0 & 12^2/2 & 12 \end{bmatrix} = \begin{bmatrix} 0 & 0 & 0 \\ 0 & 576 & 72 \\ 0 & 72 & 12 \end{bmatrix}$$

In system coordinates the flexibility matrix of the member therefore becomes

$$\mathbf{F}^1_{22} = \mathbf{T}^T_{21}\mathbf{f}^1_{22}\mathbf{T}_{21} \tag{8.52}$$

$$= \begin{bmatrix} 0 & 1 & 0 \\ -1 & 0 & 0 \\ 0 & 0 & 1 \end{bmatrix} \begin{bmatrix} 0 & 0 & 0 \\ 0 & 576 & 72 \\ 0 & 72 & 12 \end{bmatrix} \begin{bmatrix} 0 & -1 & 0 \\ 1 & 0 & 0 \\ 0 & 0 & 1 \end{bmatrix} = \begin{bmatrix} 576 & 0 & 72 \\ 0 & 0 & 0 \\ 72 & 0 & 12 \end{bmatrix}$$

The same result could be obtained by considering $\mathbf{f}^1_{22}$ as a 2×2 matrix, but in this case the transformation matrix must then be degenerated into a rectangular 2×3 matrix by dropping the row equivalent to the row and column dropped in $\mathbf{f}^1_{22}$. Equation (8.52) could thus be rewritten as

$$\mathbf{F}^1_{22} = \begin{bmatrix} 1 & 0 \\ 0 & 0 \\ 0 & 1 \end{bmatrix} \begin{bmatrix} 576 & 72 \\ 72 & 12 \end{bmatrix} \begin{bmatrix} 1 & 0 & 0 \\ 0 & 0 & 1 \end{bmatrix} = \begin{bmatrix} 576 & 0 & 72 \\ 0 & 0 & 0 \\ 72 & 0 & 12 \end{bmatrix}$$

The system-coordinate flexibility matrices of the other members similarly become

$$\mathbf{F}^2_{33} = \begin{bmatrix} 0{\cdot}6 & 0 \\ 0{\cdot}8 & 0 \\ 0 & 1 \end{bmatrix} \begin{bmatrix} 2667 & 200 \\ 200 & 20 \end{bmatrix} \begin{bmatrix} 0{\cdot}6 & 0{\cdot}8 & 0 \\ 0 & 0 & 1 \end{bmatrix} = \begin{bmatrix} 960 & 1280 & 120 \\ 1280 & 1712 & 160 \\ 120 & 160 & 20 \end{bmatrix}$$

$$\mathbf{F}^3_{44} = \begin{bmatrix} -0{\cdot}6 & 0 \\ 0{\cdot}8 & 0 \\ 0 & 1 \end{bmatrix} \begin{bmatrix} 2667 & 200 \\ 200 & 20 \end{bmatrix} \begin{bmatrix} -0{\cdot}6 & 0{\cdot}8 & 0 \\ 0 & 0 & 1 \end{bmatrix}$$

$$= \begin{bmatrix} 960 & -1280 & -120 \\ -1280 & 1712 & 160 \\ -120 & 160 & 20 \end{bmatrix}$$

$$\mathbf{F}^4_{55} = \begin{bmatrix} -1 & 0 \\ 0 & 0 \\ 0 & 1 \end{bmatrix} \begin{bmatrix} 576 & 72 \\ 72 & 12 \end{bmatrix} \begin{bmatrix} -1 & 0 & 0 \\ 0 & 0 & 1 \end{bmatrix} = \begin{bmatrix} 576 & 0 & -72 \\ 0 & 0 & 0 \\ -72 & 0 & 12 \end{bmatrix}$$

The overall flexibility matrix at point 5 must now be found: it is, of course, the sum of the separate flexibility effects at 5 due to each member in turn. The flexibility matrix at 5 due to, say, member (r) is obtained by the use of the appropriate translation matrix in the form

$$\mathbf{F}_{55}^{r} = \mathbf{H}_{i5}^{T}\mathbf{F}_{ii}^{r}\mathbf{H}_{i5}$$

For members (1), (2) and (3) the flexibility matrices at 5 therefore become

$$\mathbf{F}_{55}^{1} = \begin{bmatrix} 1 & 0 & -12 \\ 0 & 1 & 32 \\ 0 & 0 & 1 \end{bmatrix} \begin{bmatrix} 576 & 0 & 72 \\ 0 & 0 & 0 \\ 72 & 0 & 12 \end{bmatrix} \begin{bmatrix} 1 & 0 & 0 \\ 0 & 1 & 0 \\ -12 & 32 & 1 \end{bmatrix}$$

$$= \begin{bmatrix} 576 & -2304 & -72 \\ -2304 & 12300 & 384 \\ -72 & 384 & 12 \end{bmatrix}$$

$$\mathbf{F}_{55}^{2} = \begin{bmatrix} 1 & 0 & -24 \\ 0 & 1 & 16 \\ 0 & 0 & 1 \end{bmatrix} \begin{bmatrix} 960 & 1280 & 120 \\ 1280 & 1712 & 160 \\ 120 & 160 & 20 \end{bmatrix} \begin{bmatrix} 1 & 0 & 0 \\ 0 & 1 & 0 \\ -24 & 16 & 1 \end{bmatrix}$$

$$= \begin{bmatrix} 6720 & -8320 & -360 \\ -8320 & 11950 & 480 \\ -360 & 480 & 20 \end{bmatrix}$$

$$\mathbf{F}_{55}^{3} = \begin{bmatrix} 1 & 0 & -12 \\ 0 & 1 & 0 \\ 0 & 0 & 1 \end{bmatrix} \begin{bmatrix} 960 & -1280 & -120 \\ -1280 & 1712 & 160 \\ -120 & 160 & 20 \end{bmatrix} \begin{bmatrix} 1 & 0 & 0 \\ 0 & 1 & 0 \\ -12 & 0 & 1 \end{bmatrix}$$

$$= \begin{bmatrix} 6720 & -3200 & -360 \\ -3200 & 1712 & 160 \\ -360 & 160 & 20 \end{bmatrix}$$

and finally

$$\mathbf{F}_{55} = \mathbf{F}_{55}^{1} + \mathbf{F}_{55}^{2} + \mathbf{F}_{55}^{3} + \mathbf{F}_{55}^{4}$$

$$= \begin{bmatrix} 14590 & -13820 & -862 \\ -13820 & 25960 & 1020 \\ -862 & 1020 & 64 \end{bmatrix}$$

The deflection of 5 due to the load of 10 applied at 2 is also required; that is, $\mathbf{F}_{52}$ must be found. The matrix is given in this case by

$$\mathbf{F}_{52} = \mathbf{F}^1_{52} = \mathbf{H}^T_{25}\mathbf{F}^1_{22}$$

$$= \begin{bmatrix} 1 & 0 & -12 \\ 0 & 1 & 32 \\ 0 & 0 & 1 \end{bmatrix} \begin{bmatrix} 576 & 0 & 72 \\ 0 & 0 & 0 \\ 72 & 0 & 12 \end{bmatrix} = \begin{bmatrix} -288 & 0 & -72 \\ 2304 & 0 & 384 \\ 72 & 0 & 12 \end{bmatrix}$$

The load vector at 2 is

$$\mathbf{P}_2 = \begin{Bmatrix} 10 \\ 0 \\ 0 \end{Bmatrix}$$

so that the displacements at 5 due to $\mathbf{P}_2$ are

$$\mathbf{\Delta}_5 = \begin{bmatrix} -288 & 0 & -72 \\ 2304 & 0 & 384 \\ 72 & 0 & 12 \end{bmatrix} \begin{Bmatrix} 10 \\ 0 \\ 0 \end{Bmatrix} = \begin{Bmatrix} -2880 \\ 23040 \\ 720 \end{Bmatrix}$$

If the reactant forces at 5 are called $\mathbf{R}_5$, then the compatibility equations are

$$\mathbf{F}_{55}\mathbf{R}_5 + \mathbf{\Delta}_5 = \mathbf{0}$$

or

$$\begin{bmatrix} 14590 & -13820 & -862 \\ -13820 & 25960 & 1020 \\ -862 & 1020 & 64 \end{bmatrix} \begin{Bmatrix} P_5 \\ Q_5 \\ N_5 \end{Bmatrix} + \begin{Bmatrix} -2880 \\ 23040 \\ 720 \end{Bmatrix} = \mathbf{0}$$

from which

$$\mathbf{R}_5 = \begin{Bmatrix} -1{\cdot}345 \\ -1{\cdot}212 \\ -23{\cdot}6 \end{Bmatrix}$$

The matrix force method of analysis is not, of course, restricted to choices of redundants which produce tree-type statically determinate structures: this is fortunate, for such choices, while leading to a simple formulation of the problem, are usually rather bad choices from the point of view of producing a well-conditioned set of equations. A more general statement of the matrix-force method may be made as follows.

Any statically indeterminate structure may be thought of as a statically determinate structure acted on by two sets of forces: a set of externally applied loads **P**, and a set of internal redundant forces **R**. Because the structure at this stage is statically determinate, the member forces **p** for all members of the structure may be found in terms of **P** and **R**; that is, we could write down the equation

$$\mathbf{p} = \mathbf{B}\begin{Bmatrix}\mathbf{P}\\\mathbf{R}\end{Bmatrix} = [\mathbf{B}_\alpha \mathbf{B}_\beta]\begin{Bmatrix}\mathbf{P}\\\mathbf{R}\end{Bmatrix}$$

$$= \mathbf{B}_\alpha\mathbf{P} + \mathbf{B}_\beta\mathbf{R} \tag{8.53}$$

where $\mathbf{B}_\alpha$ and $\mathbf{B}_\beta$ are found by statics. The vector **p** in member coordinates is similar to the system-coordinate vector $\bar{\mathbf{P}}$ of equation (8.29) except that it is only half as large, for it represents the member forces at one end only of each member.

For any member (r) joining nodes i and j, the relationship

$$\boldsymbol{\delta}_{ij} = \mathbf{f}_{ii}^r\mathbf{p}_{ij} \tag{8.54}$$

may be obtained if the member is fixed at j and free at i. The double subscript on the displacement vector follows the convention stated in Section 8.2, so that $\boldsymbol{\delta}_{ij}$ is the deformation, or relative displacement, of i with regard to j, rather than the absolute displacement of i. A similar equation may be written for each member, and all such equations may be assembled together into the form

$$\begin{Bmatrix}\boldsymbol{\delta}_{ij}\\\vdots\\\boldsymbol{\delta}_{mn}\end{Bmatrix} = \begin{bmatrix}\mathbf{f}_{ii}^r & & \\ & \ddots & \\ & & \mathbf{f}_{mm}^t\end{bmatrix}\begin{Bmatrix}\mathbf{p}_{ij}\\\vdots\\\mathbf{p}_{mn}\end{Bmatrix} \tag{8.55}$$

or

$$\boldsymbol{\delta} = \mathbf{f}\mathbf{p} \tag{8.56}$$

where **f** is a diagonally partitioned matrix of member flexibility matrices. These equations are analogous to equations (8.28) and (8.29). Using equation (8.53) the member deformations could also be written as

$$\boldsymbol{\delta} = \mathbf{f}[\mathbf{B}_\alpha\mathbf{B}_\beta]\begin{Bmatrix}\mathbf{P}\\\mathbf{R}\end{Bmatrix} \tag{8.57}$$

Now, the applied loads and redundant forces will cause not only member deformations $\boldsymbol{\delta}$ and nodal displacements $\boldsymbol{\Delta}$, but also relative displacements between the ends of the releases where the redundants have been chosen; the rector of these release-displacements will be called ∂. By application of the principle of virtual work, equation (8.53) produces the complementary (contragredient) relationship

$$\begin{Bmatrix} \mathbf{\Delta} \\ \partial \end{Bmatrix} = \begin{bmatrix} \mathbf{B}_\alpha^T \\ \mathbf{B}_\beta^T \end{bmatrix} \boldsymbol{\delta} \tag{8.58}$$

Hence from (8.57) we may write

$$\begin{Bmatrix} \mathbf{\Delta} \\ \partial \end{Bmatrix} = \begin{bmatrix} \mathbf{B}_\alpha^T \\ \mathbf{B}_\beta^T \end{bmatrix} \mathbf{f}\, [\mathbf{B}_\alpha \mathbf{B}_\beta] \begin{Bmatrix} \mathbf{P} \\ \mathbf{R} \end{Bmatrix} \tag{8.59}$$

The compatibility equations necessary for the solution of (8.59) are thus

$$\partial = \mathbf{B}_\beta^T \mathbf{f} [\mathbf{B}_\alpha \mathbf{B}_\beta] \begin{Bmatrix} \mathbf{P} \\ \mathbf{R} \end{Bmatrix} \tag{8.60}$$

which may be written in the form

$$\mathbf{B}_\beta^T \mathbf{f} \mathbf{B}_\beta \mathbf{R} = \partial - \mathbf{B}_\beta^T \mathbf{f} \mathbf{B}_\alpha \mathbf{P} \tag{8.61}$$

which may be solved for the redundant forces **R**. It can be seen that ∂ represents the initial lack of fit in the structure (or thermal strain effects) which will in most cases be zero. Once **R** has been found, the member forces may be found from (8.53), and if the nodal deflections of the structure are required they may be obtained from the other part of equation (8.59), that is, from

$$\mathbf{\Delta} = \mathbf{B}_\alpha^T \mathbf{f} [\mathbf{B}_\alpha \mathbf{B}_\beta] \begin{Bmatrix} \mathbf{P} \\ \mathbf{R} \end{Bmatrix} \tag{8.62}$$

The close similarity in form between the general statements of the matrix-force and matrix-displacement methods should be noted.

PROBLEMS

8.1 An elastic structure consists of two elements 2–1 and 3–2 in series, as shown in fig. 8.13. The coordinates of the three points are given by the following table:

Point	Coordinates		
	X	Y	Z
1	0	0	0
2	3	4	0
3	7	6	4

The z-axis of the member coordinates of each member is in the XZ-plane. If point 1 is built-in, the flexibility matrix of the combined structure for forces and moments applied at 3 may be written in system coordinates in the form

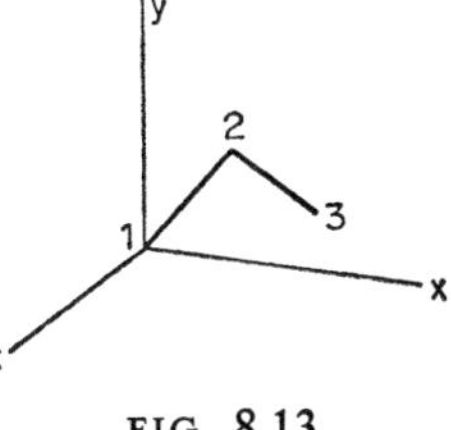

FIG. 8.13

$$\mathbf{F} = \mathbf{G}_{21}^T\mathbf{f}_{21}\mathbf{G}_{21} + \mathbf{G}_{32}^T\mathbf{f}_{32}\mathbf{G}_{32}$$

where $\mathbf{f}_{ij}$ is the flexibility matrix of a member in member coordinates and $\mathbf{G}_{ij}$ is a product of translation and rotation matrices. Calculate $\mathbf{G}_{21}$ and $\mathbf{G}_{32}$ and also the translation and rotation matrices from which they are formed.

8.2 Use the matrix force method to solve the structure shown in fig. 8.14. Neglect axial deformations, and take the two reactions at 5 as unknowns. All members are prismatic and have the same section properties.

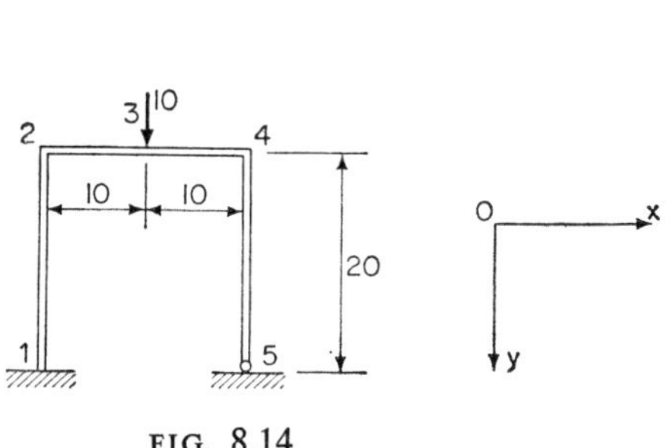

FIG. 8.14

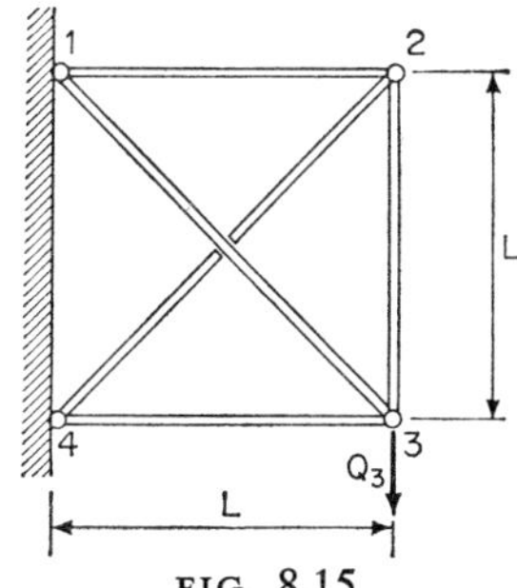

FIG. 8.15

8.3 All members of the pin-jointed truss of fig. 8.15 are uniform and have the same section stiffness EA. Use the matrix force method to find all bar forces, and also determine the nodal displacements. Choose the force in member 2–4 as the redundant.

Answer: Force in 2–1 $= +0{\cdot}442Q_3$

8.4 Do Problem 8.3 by the matrix displacement method.

8.5 Solve the structure of fig. 8.16 by either the matrix force or the matrix displacement method, taking symmetry into account in formulating the problem. All members are uniform and have the same section stiffness.

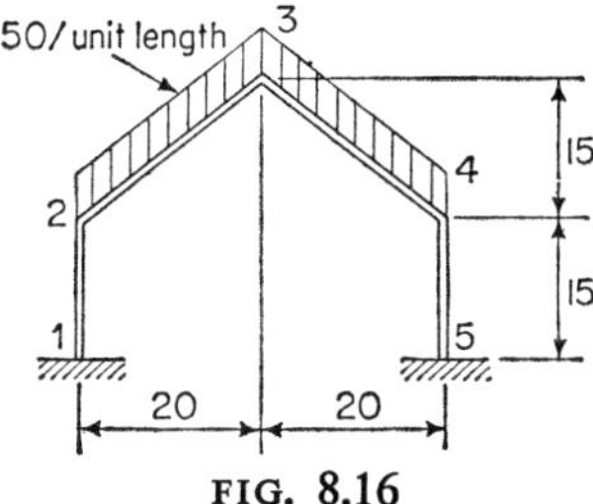

FIG. 8.16

Index